TASTY

THE ART AND SCIENCE OF WHAT WE EAT

JOHN McQUAID

SCRIBNER

New York   London   Toronto   Sydney   New Delhi

Scribner
A Division of Simon & Schuster, Inc.
1230 Avenue of the Americas
New York, NY 10020

Copyright © 2015 by John McQuaid

First Scribner hardcover edition January 2015

SCRIBNER and design are registered trademarks of The Gale Group, Inc.,
used under license by Simon & Schuster, Inc., the publisher of this work.

For information about special discounts for bulk purchases,
please contact Simon & Schuster Special Sales at 1-866-506-1949
or business@simonandschuster.com.

The Simon & Schuster Speakers Bureau can bring authors to your live event.
For more information or to book an event, contact the Simon & Schuster Speakers Bureau
at 1-866-248-3049 or visit our website at www.simonspeakers.com.

Jacket design by Will Staehle
Jacket photograph © Shutterstock

Manufactured in the United States of America

1   3   5   7   9   10   8   6   4   2

Library of Congress Cataloging-in-Publication Data is available.

ISBN 978-1-4516-8500-8
ISBN 978-1-4516-8502-2 (ebook)

For Mom

Contents

TASTY

CHAPTER 1

The Tongue Map

Early in his psychology career, Edwin Garrigues Boring often used himself as a guinea pig. As a graduate student at Cornell University in 1914, he swallowed feeding tubes to measure how his esophagus and stomach responded to different foods, and sliced a nerve in his own forearm in order to document its gradual regrowth. In 1922, just before Boring was to start a teaching job at Harvard, he was struck by a car on a rainy night. He lay in a hospital bed for six weeks with a fractured skull and short-term memory loss, forgetting his conversations with visitors within a few minutes. After he recovered, Boring used this experience to analyze the nature of awareness, pondering whether someone living in an eternal present was truly conscious.

This hands-on sensibility helped make Boring one of the twentieth century's most influential psychologists. It wasn't by virtue of any single theory or discovery. (Though he did popularize a minor curiosity, the "Boring figure," an optical illusion in which a slight shift in perspective flips the image of an old woman's face, as perceived by the eye and the mind, to that of a young woman's head.) Instead, Boring made his mark by changing the popular conception of psychology itself. When his career began, the field was a hodge-

1

podge of disciplines, equal parts philosophy, therapy, and lab experimentation, each with its own approach and terminology. From his influential perch at Harvard, Boring pushed to make it more consistent and rigorous, to have it hew more closely to the scientific method. He believed a scientist was obligated to relentlessly scrutinize and measure his own sensations, grounding all findings in direct observation—a tenet of the philosophy known as positivism. This was the closest science could possibly get to the truths about reality it aspired to capture.

But there was a point in his career when putting these beliefs into practice could have averted a major scientific misunderstanding, and Boring failed spectacularly. The mishap involved the nature of taste. By the 1940s, Boring had become an accomplished historian, chronicling the emergence and evolution of modern psychology. His 1942 volume, *Sensation and Perception in the History of Experimental Psychology*, is still considered a magisterial survey of the science of the human senses stretching back to Sir Isaac Newton's seventeenth-century studies of light and color.

Boring covered taste and smell in a relatively brief chapter in the book, twenty-five pages out of seven hundred. Midway through it, he reviewed an experiment done in 1901 by David P. Hänig, a German scientist. Hänig had brushed sweet, salty, bitter, and sour solutions—representing the four basic tastes, important components of flavor—on different areas of the tongues of volunteers, and then asked them to rate their relative strength. He found the threshold for detecting each taste varied around the edge of the tongue. The tip, for example, was more sensitive to sweetness and to salt than was the base.

It wasn't clear what this meant—if anything—and the

differences were very small. But Boring found this notion interesting and went to some lengths to illustrate it. He borrowed the data from Hänig's study and turned it into a graph. The graph was just a visual aid; it had no units, and its curves were impressionistic. But the result was that—perhaps to dramatize the point, or perhaps inadvertently—Boring made small differences in perception appear huge.

The wayward chart became the basis for a famous diagram of the tongue, divided into zones for each taste: The tip is labeled sweet and the back bitter. Along each side, salty is close to the front, and sour is behind it. The center is blank. Linda Bartoshuk, a professor of psychology who has studied this map's origins, believes it came about through a game of "telephone": First, Boring exaggerated Hänig's findings. Then researchers and textbook editors misinterpreted Boring's graph, using the peaks of its curves to label specific areas on the tongue. A final round of confusion produced a diagram with taste boundaries clearer than those on a world map.

The tongue map offered a simple explanation for how the tongue processed tastes, a phenomenon everyone knew intimately. Teachers embraced it. Generations of elementary school students sipped and swished water spiked with either sugar, salt, lemon juice, or tonic water in a classroom experiment designed to dramatize the tongue map. Like air raid drills or dodgeball, the tongue map became a feature of postwar American schooling and lodged itself in the popular imagination.

However, these demonstrations no doubt confused more children than they enlightened, as many found they couldn't detect the supposedly dramatic taste gradients. Even as the tongue map took on the mantle of conventional wisdom, research revealed that it wasn't merely an exaggeration or misinterpretation but totally wrong. In 1973, Virginia Collings of the University of Pittsburgh repeated Hänig's original tests. Like him, she found very limited variation in the tongue's taste geography. In the 2000s, more advanced tests showed that all five tastes ("umami," or savoriness, was recognized as a fifth in 2001) can be detected all over the tongue. Every taste bud is studded with five different receptor proteins, each tailored to detect molecules of one of the basic tastes.

Had Boring done some taste testing himself instead of interpreting Hänig's forty-year-old data, he might have noticed the problem with his graph. Instead, he launched one of the more widely disseminated bits of scientific misinformation in history.

The old diagram has lost much of its cachet in recent years. But it still lingers in some areas of the culinary world, including coffee and wine tasting, which value tradition and continuity as much as science. Claus Riedel, the Austrian glassware designer, used it to create wineglasses whose unique curvature is intended to deliver each sip to the right place on the tongue to release the full flavor. (Riedel died in 2004; since then, his son and successor Georg Riedel has acknowledged that science undermines the tongue map, but maintains the glass designs work.) Boring died in 1968, before the map had been discredited. That he made a fundamental error about the nature of one of the senses, which he considered the building blocks for understanding both the mind and the universe, is an irony that he would doubtless have found mortifying. It was no mere miscalculation, but a basic error about a universal human experience. Everyone knows the gratifying "Mmm" of sweetness and the stark taste difference between a pinch of salt and a fistful. Cheesecake makes the brain explode with pleasure. The complex tastes in coffee are a global obsession. Recipes distill entire cultures down to a single sensation. Flavor is one of a very few things that make day-to-day existence not just survivable but consistently enjoyable.

Why did this happen? Mere carelessness doesn't seem like an adequate explanation of Edwin Boring's disregard for his positivist philosophy, the foundation of his life's work; nor does the fun of the tongue map experiment completely account for its persistence given that it never really worked. Boring's mistake may be flavor's version of the Freudian slip, an apparently superficial error that reflects hidden conflicts.

One reason for the befuddlement is that for thousands of years, scientists and philosophers have viewed taste and flavor as less-than-worthy subjects for study. The ancient Greeks considered taste to be the lowest and grossest of the senses. Vision can discern the subtleties of high art or the smile of a loved one, but taste's mission is simple: to distinguish food from everything else. The Greeks thought the temptations it posed in carrying out that mission clouded the mind. In his dialogue *Timaeus*, Plato wrote that the sense of taste was caused by the varying roughness or smoothness of "earthy particles entering into the small veins of the tongue, which reach to the heart." The heart was the seat of the baser bodily sensations, while thought and reason occupied the "council chamber" of the brain. Of course, food was headed for the belly, a ravenous beast knocking about far below the deliberating council: "the belly would not listen to reason, and was under the power of idols and fancies."

Plato practiced what he preached: in his *Symposium*, guests gather for a banquet, but decline food or drink in order to keep their minds clear for the discussion on the nature of love.

These biases became a fixture in thought about the senses

for centuries. The German philosopher Immanuel Kant wrote in the eighteenth century that flavor was too idiosyncratic to be worthy of study. To Kant's eye there were apparently no universal principles governing it, like those governing the behavior of light. Even if there were, they could not be derived from observation, because there was no way to observe the mind. Taste would always elude us. Kant's contemporary David Hume disagreed, arguing that good taste in food was tied to good taste in art and all things. But it was Kant's more skeptical attitude that persisted.

These withering assessments overlook a great deal, and they reflect a certain discomfort. Flavor embodies the basic savagery of being an animal, devouring the flesh of other animals and plants in order to stay alive—and loving it. In flavor the order of civilization momentarily disappears and is replaced by carnage. Confronting this part of human nature is unnerving. Eating and drinking are also forms of intimacy that are, in their own way, as powerful and unsettling as sex; after all, they involve taking something inside the body, many times a day, with flavor as a seducer. Flavor is a conscious manifestation of ancient, inexorable drives that make life possible. Sigmund Freud believed the central drama of life sprang from the sex drive. But the drive for sustenance, which runs on a similar cycle of craving, pleasure, release, and satisfaction, maintains a more powerful, consistent grip on our lives and motivations.

The other problem in studying flavor is the basic inscrutability of a phenomenon that unfolds entirely within the body, brain, and mind. Vision, hearing, and touch are "shared" senses. We all see (or think we see) the same shades of color, hear the same sounds, feel the same textures with our fingertips. This gives scientists a common frame of reference with which to conduct experiments, collect data, and compare notes on these phenomena and the senses that perceive them.

There is no such shared reality with flavor. Like light and sound, the chemical components in food and drink are objective, measurable quantities. Yet the perceptions of them vary wildly from person to person: There are delicate sensibilities and dull ones. Foods beloved by some are despised by others. Tastes vary by culture, by geography, even by one's mood. A scene from *Don Quixote* captures these subtleties. Sancho Panza, the slovenly, loyal squire, boasts to strangers that a sensitive palate (a sign of good breeding) runs in his family. He tells the tale of two relatives judging a fine wine in a tavern. One sips it, swishes it in his mouth, and says it's wonderful, except for the slight taste of leather. The other takes a drink. It's excellent, he says, except for that off hint of iron. The barflies mock the relatives for putting on airs. But when the wine cask is emptied later, the tavern owner finds an iron key on a leather thong.

Such differences of perception offer hints at flavor's inner workings, a secret world lying just under the surface of everyday experience, waiting to be cracked open. But this subjectivity is also exactly what makes it so difficult to formulate general principles about flavor chemistry or perception. Newton devoted years to the study of light and color perception, and founded the science of optics—demonstrating, among other things, that white light was not the absence of color but a blending of all colors. But there was never a Newton of taste—no Enlightenment-era scientist to revolutionize the field and set it on a path to a modern understanding.

The combination of inaccessibility and unease conspired to keep taste and flavor on the scientific margins for most of the past two thousand years. The ancient Greeks first formulated the idea of "basic" tastes as irreducible elements, the atoms of flavor. Some of the earliest known attempts to explain taste were made by the physician Alcmaeon, who lived in the Greek city of Croton in Italy and wrote sometime between 500 and 450 BC. He thought the tongue, like the eyes (and nose and ears; for some reason, touch was left out), had its own *poroi*, or channels, that transmitted perceptions, like barges delivering amphorae of wine, to the brain. That's exactly what nerves do. In the fifth century BC, the philosopher Democritus declared that the perception of taste depended on the shape of individual atoms, the postulated smallest units of matter: sweet atoms were round and relatively large, so they rolled around on the tongue, salt atoms were shaped like isosceles triangles, and pungent ones "spherical, thin, angular, and curving," prone to tear the tongue's surface and generate heat through friction, which explained the irritation they caused.

With only minor variations, this conception of flavor has been a staple in most societies and civilizations from then to now. Traditional Indian medicine, known as Ayurveda, Sanskrit for "life-knowledge," employs combinations of sweet, sour, pungent, bitter, salty, and astringent tastes to combat illnesses. Its weight-loss diet prescribes foods that are pungent (a product of the elements fire and air) or bitter (air and ether) to combat an excess of *kapha*, or phlegm (earth and water). Carl Linnaeus, the Swedish botanist who in the eighteenth century invented the modern scientific system for naming species and classifying life, identified the basic tastes as sweet, acid, bitter, saline, astringent, sharp, viscous, fatty, insipid, aqueous, and nauseous. The idea behind the tongue map, that nature has cleanly divided the tongue's territory into taste regions, grows out of this tradition. It is simple and appealing, like the nineteenth-century phrenology diagrams that mapped various mental capacities onto areas of the skull. But in recent years, flavor's once mysterious, closed-off domains have begun to open up. Scientists employing new tools and technologies have deepened our understanding of flavor, helping it to shake off its second-class status among the sensory phenomena and placing it squarely in the vanguard of the study of human biology.

The old philosopher's claim that flavor was impervious to scientific inquiry is now moot. Flavor science made great strides in the twentieth century; in the twenty-first it has moved ahead at astonishing speed. Receptors for all five basic tastes have been found, and it looks like fat may be identified as the sixth. Scientists are beginning to understand the connections between mind, brain, and body: why you think you *have* to have that cheeseburger or glass of wine.

This book is a brief biography of flavor. The narrative begins at the dawn of life on earth and ends in the present, and explores the structure of this unique sensation, from its molecular building blocks through more sophisticated levels of body, brain, and mind. Flavor has grown deeper and more complicated at each stage in its development over millions of years. It has driven evolution, and lately human culture and society, in new directions. It is a kind of slate on which human struggles, aspirations, and failures have been written, erased, and written again. We owe our existence and our humanity to it—and, in many ways, our future depends on it, too. As science unlocked flavor's secrets, its influence over what we eat and drink exploded. From the food labs of large corporations to the kitchens of the world's finest restaurants to the bar down the street, science shapes surprising and sometimes alarming new sensations tied to both our DNA and our deepest drives and feelings.

In March 1998, scientists at the National Institutes of Health (NIH) in Bethesda, Maryland, found themselves on the cusp of one of these paradigm-changing advances. They were searching for a sweet taste receptor, a kind of protein on the tongue specially tailored to snare sugar molecules out of the slurry of mashed-up food and drink in the mouth. More than two thousand years after Democritus and Alcmaeon, science was finally closing in on the mechanism of taste that enables us to transform the molecular arrangements in food into sensory perceptions and, ultimately, culinary art.

Over the previous decade, the science of genetics had made startling advances. For the first time, scientists were decoding the entire length of human DNA, the helical, ladderlike molecules found in chromosomes in the nucleus of every cell. The total human genome is made up of three billion pairings of four amino acids; each pair forms a rung in the ladder. The variations in the pairs constitute a code that maps out the blueprint for the body and all its functions. Every person gets two of these blueprints, one from each parent. Uncoiled, the DNA in a single cell would be about six feet long; if all the DNA in the human body were laid end to end, its length would be the equivalent of seventy round-trips from Earth to the sun.

Isolating genes—discrete segments of DNA that carry out specific biological instructions such as making proteins, the body's basic building blocks—had enabled scientists to find and treat diseases and to better understand human evolution. Now, genetics offered a way to quantify the intangibles of flavor, to explain its bewildering diversity. The nose's receptors for smell had been isolated and their genes decoded seven years earlier, an effort that later won a Nobel Prize. The smell receptors had been comparatively easy to find. They were plentiful and concentrated in the small patch of tissue on the roof of the nasal cavity; live ones can be harvested with a Q-tip.

But the search for taste receptors had dragged on. They had proven almost impossible to isolate: Not only are there relatively few taste-detecting cells to begin with, but it is difficult to coax a reaction from them. The body has a vast apparatus to detect all kinds of cues, from hormones on the inside to heat, cold, pressure, light, and chemicals on the outside. Most of these reactions are very sensitive. It takes only a tiny dose of adrenaline to get a rise out of the receptors that detect it. But taste receptors are about a hundred thousand times less sensitive. This is because they interface with the chaos of the world around us. Given the sheer volume and variety of sensations the tongue encounters in a single meal, the brain would overload if every molecule lit up the taste receptors. Taking a sip of Coke would be like staring into the sun.

The NIH scientists, led by Nick Ryba, had finally leaped many of those hurdles. They were examining taste bud cells while also searching stretches of the genome, hoping to match up a taste receptor protein with the gene that created it. They harvested DNA from the taste buds of rats and mice, whose sense of taste is similar to our own. The trick was finding the right individual gene: a short, specific stretch of code tucked somewhere among vast, unmapped tracts of DNA. Having that blueprint would enable them to clone copies of the receptor so they could easily study its structure and workings.

In a short span of time, science had become quite adept at slicing, dicing, and sorting these once-indistinguishable molecular strands. At NIH, scientists found a way to turn the scarcity of taste receptors to their advantage: They used a technique that plucked only the most unusual DNA snippets from across the tongue, separating them from more generic strings. Some of these had to contain the material for taste. Next, they took each fragment and injected it into taste cells harvested from rodents. If it latched onto the DNA in the cell, it was a taste receptor gene. This was, roughly speaking, like putting a toddler in a room with a woman you think may be the mother: if they hug, then you know they're related.

It worked: the scientists found half of a rodent gene for a sweet receptor; the second half was found soon after, and then the analogous human gene for sweetness. Their double genes mean that sweet receptor molecules come in two parts that fit together like a train coupling. They are bizarre, Lovecraftian-looking things, tangled skeins of seven coils stuck in the surface of a taste cell. One coil reaches out into the void to snag sugar molecules floating by. When it does, an electrochemical chain reaction begins that travels all the way to the brain, igniting a burst of pleasure.

Elsewhere, scientists were starting to crack another once-intractable problem, the subjectivity of taste. A few years after the sweet receptor discovery, volunteers in an experiment at the University of Groningen in the Netherlands lay on a table with pacifiers connected to long straws in their mouths. They were then slid into a magnetic resonance imaging (MRI) scanner that recorded their brain activity as they sipped bitter tonic water through the straws. Later, they were scanned while looking at photos of people grimacing in response to a taste of a drink, and again as they read brief scripts intended to evoke distaste or disgust. The purpose of this experiment, run by neuroscientist Christian Keysers, was to explore the relationship between tastes and emotions. During the 1990s, the emergence of this type of scanner, called a functional MRI (fMRI), allowed scientists to see which parts of the brain were active when a person ate or drank, smelled an aroma, or read—anything that could be done while one's head was immobilized.

There were limits to this approach. It showed associations between real-world actions and arcing networks of neurons in the brain, but not exactly what those associations meant. But it was a revealing waypoint between the chemical reaction of taste on the tongue and the mind itself.

Their findings were strange. As volunteers imagined bitterness in a story, or saw photos of a wince of distaste, their brains experienced a "bitter" reaction. These patterns varied slightly in each part of the experiment, reaching out to encompass different parts of the brain. Taste seemed to be a cornerstone of higher functions such as imagination and emotion.

The next twist in this story is still being written; it hinges on certain lingering mysteries. Flavor remains frustratingly paradoxical. Like other senses, it's programmed by genes; unlike them, it is also protean, molded by experience and social cues, changing over the course of a lifetime. This plasticity is wild and unpredictable: people can learn to like or dislike almost anything, which is why the range of flavors in the world is seemingly infinite, and why the old tongue map was useless.

Everyone lives in his or her own flavor world, which takes form during early childhood and evolves over the course of our lives. This world is created by the clash of ancient evolutionary imperatives meeting a lifetime's worth of high-octane processed foods, cultural cues, and commercial messages.

The flavor preferences of my own children, born two years apart, were apparent as soon as they began to eat solid food. Matthew, the elder, relished extremes. He started eating jalapeños in preschool and liked coffee from the time he was nine.

Every so often, usually in the summer, he would sit down with a lemon or lime, quarter the fruit, add salt, and devour it with the peel. His sister, Hannah, craved bland, rich flavors, and the foods she ate tended to be white or beige: cheese, rice, potatoes, pasta, chicken. She preferred chamomile tea to coffee, and milk chocolate to dark. Yet both were picky eaters: they knew what they liked and rarely departed from it. Getting them out of their respective comfort zones to try something new was nearly impossible.

This combination of divergent tastes and limited likes turned grocery shopping or restaurant-going into a kind of Rubik's cube challenge; only pizza satisfied everyone. I made most family dinners, and struggled to get them out of a rut dominated by the same handful of dishes presented with only slight variations in a weekly cycle: pasta, roast chicken, or chicken nuggets for Hannah, hot dogs or shrimp in Szechuan sauce for Matthew. My wife, Trish, and I were more adventurous, but the convenience of this routine dragged us in, too.

The appetites of children are a crucible where the forces of chemistry and culture collide. The sweet tooth, the scourge of modern nutrition and dentistry, is crucial to childhood development. In newborn babies, sugar acts like aspirin, soothing pain. The Monell Chemical Senses Center in Philadelphia, a think tank that studies taste and smell, found that children with a strong taste for sweets also had higher levels of a hormone tied to bone growth. A yearning for sweets pointed early human children to then precious sugars in fruits and honey, and when combined with sourness, to citrus fruits packed with vitamins C and D.

Picky eating is likely a holdover from the same epoch, when humans lived together in small migratory groups and children—thanks to their tendencies to wander and to shove random things in their mouths—faced a constant threat of poisoning. Today, a limited diet is a danger to long-term health, and in its most extreme form pickiness has been labeled an eating disorder, called food neophobia.

Children have strange tastes because they are bizarre creatures. Taste and smell develop earlier than other senses, so a fetus's sensory universe consists almost entirely of the smells and tastes in amniotic fluid. This makes a lasting impression. In another Monell study, the babies of women who drank a steady diet of carrot juice during their pregnancies or during breastfeeding later took a shine to carrot-flavored cereal.

Then, between birth and the ages of two and three, a baby's synapses—the connections between neurons that form networks in the brain—multiply from about 2,500 per neuron to 15,000 (an adult has 8,000 to 10,000). This temporarily ties the senses together. Young children live in a fugue of overlapping sensations, one reason why early flavor experiences evoke not just meals but entire moments. As children age, experience gradually trims the thicket of neurons, and better sensory connections emerge. During this process, kids' tastes vacillate between conservative stretches and probing, adventurous periods.

During the teen years, intense tastes fade along with the physiological demands and evolutionary imperatives of childhood. A subtler palate takes their place, though the original likes and dislikes never quite disappear. This muting allows the range of tastes we can experience to increase, and our reservoir of food memories and associations deepens. Sensations bubble up, synapse by synapse, from chemical reactions in the nose and mouth. Meanwhile, food engages the other senses, tapping the mind's capacity for learning, understanding, and appreciation. Back and forth it goes: the mind shapes taste, and experience shapes the mind. A version of this dialogue has gone on through billions of meals since life first developed an appetite.

CHAPTER 2

The Birth of Flavor
in Five Meals

The first inklings of flavor appeared as early life-forms began to sense the world around them and the taste of nutrients floating by in seawater excited primitive nervous systems. Countless meals were consumed as life evolved over the hundreds of millions of years that followed. Like Russian nesting dolls, our modern tastes contain those experiences. No matter how cultured one's palate or subtle the ingredients in a dish, a taste summons raw urges out of the deep past, echoing evolutionary twists and long-ago life-and-death struggles over food. Five ancient meals, each taking place at a turning point in evolutionary history, help explain where the sense of flavor, and *Homo sapiens'* talent for culinary invention, came from.

The First Bite

The creature resembled a scarab. About an inch long, with a soft, ribbed carapace, it scuttled across the sand in a primordial coastal shallow. Then it sensed a threadbare tapestry of

17

smells, vibration, and shifting light. Its wormlike prey bur-
rowed into the sand, trying to undulate its way to safety. But
it was too late. The predator ripped it open with its pincer-like
mandibles, sucked it into its mouth and down its gullet, then
continued on its way, searching for a sheltered spot to digest.

Evidence of this 480-million-year-old meal was discovered in 1982, when a scientist named Mark McMenamin on a
survey expedition spotted a tiny fossil imprint in a gray-green
slab of shale. Without giving it much thought, he chiseled
the impression out of the rock and bagged it along with dozens of other samples. Then a graduate student, McMenamin
was surveying the geology of the Sonoran Desert for the Mexican government, picking over the flanks of Cerro Rajón, a
summit about seventy miles southwest of Tucson in the Mexican state of Sonora. The ancient seabed had ended up on a
mountaintop.

To the untrained eye, the fossil looked like a series of
faint scratches barely a quarter-inch long. When he studied
it back at the lab, McMenamin recognized them as traces of
the movements of a trilobite, etched into petrified mud. Trilobites were the ancestors of nearly everything in the animal
kingdom: fish, flies, birds, humans. They left countless fossils
in seabeds, making them a fixture in natural history muse-
ums. Many had shells with multiple segments and looked
like a cross between a horseshoe crab and a centipede. This
fossil's pattern of markings was well-known, and even had a
scientific name, *Rusophycus multilineatus*. McMenamin kept it
and wrote about it in his PhD thesis. He thought little about
it until more than twenty years later, when he was a professor of geology at Mount Holyoke College, studying the early
evolution of life.

McMenamin was examining the fossil again when he saw

something he had previously overlooked. "It had this additional feature, not just the trilobite, but another sinuous trace fossil right next to it," he said. "These things are rare." He concluded the fossil contained evidence of an encounter between two creatures. The extra trace was an indication of a smaller, wormlike organism's attempt to burrow into the mud. From the arrangement of the markings, it appeared the trilobite had been right on top of it. McMenamin employed Occam's razor: the simplest explanation was that the trilobite had been digging for lunch. This was, he wrote, evidence of the "first bite," the oldest known fossil of a predator eating its prey.

What did this meal taste like? Is it even possible to imagine?

Before this era, known as the Cambrian Period, flavor did not exist in any meaningful sense. Life on earth consisted mostly of floating, filtering, and photosynthesis. Bacteria, yeasts, and other single-celled creatures nestled in the furrows of granite and between grains of sand. Some joined together into slimy mats of cells. Organisms shaped like tubes or disks rode the ocean's currents. "Eating" meant absorbing nutrients from the sea. Sometimes one organism enveloped another.

Then, over tens of millions of years—suddenly, in geological terms—the seas filled to teeming with new creatures, including the trilobites, which became the most successful class of organism in the history of life; their dominion lasted more than 250 million years. Their emergence, about 500 million years ago, was when nature as we know it really began: for the first time, life began systematically devouring other life. Unlike their predecessors, these new creatures had mouths and digestive tracts. They had rudimentary brains and senses that allowed them to detect light, dark, motion, and telltale chemical signatures. They used this fancy new equipment to hunt, to kill, and to feed. As Woody Allen's character Boris remarks in the film *Love and Death*: "To me, nature is . . . I dunno, spiders and bugs and big fish eating little fish. And plants eating plants and animals eating . . . It's like an enormous restaurant."

No trilobites survive today, and fossils do not reveal much about their nervous systems, so assessing their sensory capabilities depends on educated guesswork. Certainly, they could perceive nothing like the complicated flavors of dark chocolate or wine. Human tastes, even the aversive ones, are full of subtleties and associations with other flavors, and to past events and feelings, the whole of our learned experience. Trilobites probably did not feel anything like pleasure, and retained only a few trace memories. Each meal would have tasted more or less the same. Its saliency would have come mainly from the slaking of hunger and the urge to attack.

Still, these primordial elements of flavor were an extraordinary evolutionary achievement, and human tastes share this same basic physiological structure. Of course, that's something like comparing a mud hut to Chartres Cathedral. But the foundation had been set.

Some big change occurred in living conditions on earth to trigger this predator-prey revolution, which is called the Cambrian explosion. Scientists disagree about what it was. Some believe it was caused by a prehistoric bout of global warming that had melted the polar ice caps after a long deep freeze. The seas rose hundreds of feet, and water rolled far inland, over low hills and rocks with lichens and fungi (trees, grasses, and flowering plants did not yet exist), carving out lagoons and shaping sandbars and shoals, creating warm, shallow cauldrons ideal for life to flourish. Others trace it to a shift in the orientation of the earth's magnetic field, still others to mutations that brought about the emergence of the action potential, the ability of nerve cells to communicate over distances, or other fortuitous changes in the DNA code.

Whatever the precise sequence of events, an iron link was established between acute senses and evolutionary success. A biological arms race ensued as bodies and nervous systems adapted to rising threats and opportunities. The senses, once mere detection-and-response mechanisms, had to grow more powerful in order to guide complicated behavior. Flavor became the linchpin of this process. From the time of the trilobites to the present, foraging, hunting, and eating food have driven life's endless bootstrapping, culminating in our big human brains and the achievements of culture. More than vision, or hearing, or even sex, flavor is the most important ingredient at the core of what we are. It created us. The ultimate irony, McMenamin says, is that the introduction of killing into the world, and with it untold suffering, also expanded intelligence and awareness, and ultimately led to human consciousness.

Sweetbreads

Drawn by the scent of decomposition, the jawless hagfish burrows into the bodies of dead sea creatures and then devours their carcasses from the inside out. This has proven to be a highly successful evolutionary strategy. Jawless fish, the first vertebrates, appeared about 450 million years ago, roughly 30 million years after the "first bite," and the fossil record shows they have changed little since then. They are older by about 200 million years than their rival for the title of champion survivor, the cockroach. The hagfish, an outlandish-looking animal with an eel-like body and a sucker for a mouth, is sometimes called a living fossil. Humans are descended from some ancient relative of the hagfish; its anatomy and behavior offer a glimpse into the deep past, when the basic couplings between the brain and the senses were first established.

To early predators, the trilobites, taste and smell would have been virtually indistinguishable. But in jawless fish, they assumed different jobs, and would not reunite until humans appeared on the scene. Taste became a gatekeeper to the body's inner precincts. But smell reached out into the world. Hagfish swam through a shifting haze of scents. Smell created a picture of their surroundings in their brains: predators, potential mates, their next meal. To humans, the scent of rot usually triggers disgust. But this reaction is subjective. To the jawless fish, it meant survival and satisfaction.

Where did this additional sensory power come from? Sometimes, mutations in the genetic code do not merely change the body—they add to it. Entire strings of DNA can randomly duplicate themselves; when their biological instructions are carried out, the organism acquires an extra set of something. Redundant tissues can be deadly, mucking up the body's normal functions. But under the right circumstances, they can bring about significant evolutionary leaps. The original genes continue doing their established jobs, and natural selection works on the copies, which take on new tasks or build new body parts. The German writer and naturalist Johann Wolfgang von Goethe anticipated this powerful evolutionary force in the late eighteenth century, guessing that duplicate parts of the anatomy might transform themselves into something different. The structure of a leaf might be the basis for the flower petal. A skull might be a modified vertebra.

In the jawless fish, receptors for smell were duplicated and the extras altered to detect new scents. Their immediate ancestors likely had only a handful of smell receptors; hagfish have more than two dozen. As life evolved, this process repeated itself many times over: some animals have as many as 1,300 kinds of smell receptors; humans have more than 300.

The new sensations bombarding the first jawless fish would have been a cacophony to the brain of the average trilobite. So as the sense of smell grew sharper, the hagfish brain adapted. The olfactory bulb is a way station between the nose and the brain of all animals, converting smells to patterns of nerve impulses. In the hagfish, a new structure grew upward from the bulb, like a flower springing out of the earth. This structure was the forerunner of the cerebrum, the topmost part of the human brain that gives conscious form to virtually everything we do: it processes senses, perceptions, movement, and speech. In humans, the same sets of genes still jointly govern development of the olfactory anatomy and the brain's basic structure. Smell has been the biological currency of feeling and action for almost as long as animals have had nostrils. It is the human sense of smell that gives flavors their vast range and subtleties. Proust, whose novel *In Search of Lost Time* is a reverie inspired by the scent and taste of a madeleine cookie dipped in tea, might have been taken aback to hear that carrion feeding was the starting point for humanity's deep connection between smell and memory.

Ant Soufflé

About 250 million years ago, the global dinner table was abruptly cleared and reset. A wave of volcanic eruptions across the Siberian steppes, possibly triggered by a meteor impact, sent lava pouring over nearly a million square miles. Ash blotted out the sun for millennia. Acid rains raked the face of the planet. Plant life in the oceans and on land died off, and the atmosphere grew thick with carbon dioxide, making it all but unbreathable. This cataclysm, called the Permian extinction, eliminated 90 percent of marine species and 70 percent of land species (even most insects, which often escape such catastrophes). It was the biggest mass extinction in the history of life, a bookend to the Cambrian explosion 250 million years earlier.

Into this blighted landscape sauntered two quite different kinds of animals: dinosaurs, and creatures that looked something like small furry lizards. The outlines of this story are familiar: dinosaurs dominated the planet until their own end came, while early mammals stayed out of their way, waiting their turn. But in the shadows and hollows where mammals skulked, a different storyline was unfolding.

One of these protomammals, *Morganucodon oehleri*, lived about 50 million years after the Permian extinction. It wasn't cuddly; *Morganucodon* laid eggs, and its long snout and ambling gait were reptilian. It had some mammalian traits: fur, warm-bloodedness, and a secondary joint in its jaw. But what really placed *Morganucodon* closer to the mammal camp were its heightened perceptions, forged around its endless hunt for food, which became the object of complex strategies and vivid gratification—the earliest stirrings of humans' grand culinary passions.

Morganucodon was a wisp of a beast, shorter than a man's finger, but its whole body responded to the world. In a single moment, it could register the scent of a tiny lizard a hundred feet away, a termite mound over the next rise, and a dinosaur across a bog. Its eyes could spy predators in the dark. It could sense animals moving nearby via slight shifts in airflow over its fur. Whiskers helped it root through the underbrush for food. It usually found what it was looking for: trails that led to anthills, worms and grubs under rotting tree trunks, tinier mammals skittering across its path. Mealtimes, which in earlier epochs were all about filling the stomach and closing the abyss of hunger, were now focused more on the delicate senses of the mouth, offering earthy flavors and hints of pleasure.

This was the world of the scavenger. If food wasn't quickly and efficiently obtained, eaten, and digested, a *Morganucodon* would die, either of starvation or as some dinosaur's snack. Mammals' signature advance—warm-bloodedness—reflects this desperation, and the crisp urgency of each meal. Cold-blooded dinosaurs could eat and rest at varying tempos depending on how hot or cold it was, husbanding their energy. Mammals had to be constantly on the hunt, and good at it, because the metabolic furnace that maintained their body temperature demanded far more calories (a modern mammal at rest consumes seven to ten times the energy of a reptile the same size). As time went by and dinosaurs grew larger, mammals had to pour still more energy into evading them.

A new brain structure evolved to manage these challenges. In humans, the neocortex is the outer layer of gray tissue that covers the rest of the brain (*cortex* is Latin for "rind"). Only mammals have a neocortex, and most are smooth; only human and ape neocortices are lined with the characteristic grooves and folds that greatly increase surface area, and thus processing power. Structures in the neocortex are responsible for most of our conscious perceptions, including flavor. It's here where feelings, urges, and impressions bubble up to awareness and spur us to act. But the early mammalian neocortex's most important job was to be a map of lived experience, recording smells, mates, threats, and meals—what tasted good and filled the stomach, where it could be found, and the tactics that had obtained it. Flavors now made tightly woven neural patterns of sensation, memory, and behavioral strategy, constantly updated and molded by new events.

Tim Rowe, director of the vertebrate paleontology lab at the University of Texas, was investigating the emergence of the early mammalian brain when he came up against a serious problem: there was barely any evidence to examine. Brain tissue doesn't fossilize. Nor did the soft, cartilaginous skulls of many early mammals. *Morganucodon* and some later relatives had bony skulls, but the fossils they left behind were tiny, and so old they might crumble at the slightest touch. But Rowe devised a clever way around this obstacle.

In 1997, he started to use a CT scanner to create three-dimensional images of meteorites. These were crude at first, but as computing power geometrically increased in the 2000s, Rowe modeled smaller and smaller objects, focusing on early mammal fossils. He got permission to scan a *Morganucodon* skull. As with Mark McMenamin's Promethean bite, Rowe found new insights in an old fossil that had been sitting on somebody's shelf; this one occupied a lab case at Harvard, where Rowe had himself handled it twenty years earlier. Now, he gently placed it on a small table inside the scanner. It spun, and over the course of five or six hours the scanner built an image of the skull, voxel by voxel (voxels are three-dimensional pixels, the smallest components of the image). When complete, Rowe could enlarge the inch-long skull to the size of a ranch house. Studying every microscopic bump and fold in the bone, and cross-referencing it with ancient and modern anatomy, Rowe constructed a model of the brain that had occupied it, and a picture of life on the cusp of change.

The brain was 50 percent larger, relative to body size, than those of *Morganucodon*'s immediate ancestors; a sharpening, more expansive sense of smell accounted for most of the growth. Early mammals likely had more than a thousand distinct smell receptor genes, making them far more sensitive to scents than dinosaurs, which had perhaps a hundred. Rowe's work suggests this was merely the first of several large pulses of smell-brain growth. He scanned another fossil skull belonging to the species *Hadrocodium wui*, a relative of *Morganucodon* that lived about 10 million years later. (Both fossils were found in China.) *Hadrocodium*'s skull was only about a third of an inch long, broken into dozens of nearly microscopic pieces. But once scanned and virtually reassembled, it revealed a brain almost bursting with new nerves and perceptions. It was bigger overall and its neocortex more elaborate, with more power to process and weave together all of the senses. At its base, the spinal cord bulged, suggesting more complex connections between the body and the brain, and that it moved more quickly and gracefully than its predecessors.

Echoes of this epochal transition persist in the fetal development of all mammals today. The first region to develop in a mammalian fetus's neocortex is the area that represents the mouth and tongue, because of the essential role of nursing in its survival. The earliest sensations it processes are the warmth, smell, sweet flavor, and deep gratification of mother's milk. The first mammals had long snouts and powerful lips, as well as well-developed whiskers. The mouth and nose became more than just anatomical tools for tracking food; they made food a focal point for all experience. In the great scavenger hunt, the mouth and nose led the way.

Fruit Salad

It was just a flicker of orange, but it blazed through the green. The band of monkeys, living in the African jungle about 20 million years ago, had spent days chewing on duller fare: mostly leaves, bitter roots, and bugs, plus a few pungent berries. Suddenly, here was the hint of something great. As they clambered over tree branches, their eyes narrowed, and more flecks of orange appeared. They leaped, swinging in unison to the right spot. They grabbed the reddish-orange fruits with all five fingers, crushing them, letting the juice dribble over their hands. One squatted on a branch, leaned back against the tree's trunk, and bit into the fruit. The sweetness exploded on his tongue, tempered by a bitterness; a brief, blinding shock of pleasure. The feast would go on until pits littered the forest floor.

The monkeys' world would have encompassed only a few square miles, a territory probably similar in area to that of *Morganucodon*. They had also evolved in similar circumstances— scavenging and hiding from predators in the wake of the ecological catastrophe that killed off the dinosaurs, a giant meteor strike off the coast of the Yucatán Peninsula. But there were two crucial differences. Our ancestors' hunt for food, formerly a ground-based affair, had moved upward into the trees. It now occupied three dimensions instead of two, and a new form of vision, paired with depth perception, rendered it in vivid colors. This advance yoked vision closely to flavor. The bright color of the forbidden fruit must have been what first caught Eve's attention in the Garden of Eden, and this is still the case with our own meals. Colors, shapes, and the arrangement of food draw the eye and whet the appetite.

Most mammals have two-color vision: their retinas, the image-sensing area at the back of the eyeball, contain two kinds of specialized sensors called cones, with receptors that detect blue or red wavelengths of light. An animal with two-color vision can distinguish about ten thousand distinct hues. But about 23 million years ago, a gene replication occurred in a species of monkey. Those affected received a third set of cones that became tuned to the yellow band of the spectrum. Hues that had appeared flat and gray to earlier mammals now became purple, pink, sky blue, mauve, teal, coral. Reds grew deeper and subtler, greens softer and more varied. Primates with this enhanced vision, which today include some—but not all—monkey species, all apes, and humans, can detect up to a million colors. (Birds have four types of cones, and fantastically rich color vision.)

Finding fruit in a jungle setting is difficult, a "Where's Waldo?" problem: the eye and brain must detect a distinct color signal out of the predominant hue. In the 1990s, Cambridge University neuroscientists Benedict Regan and John Mollon set out to test the fruit-vision hypothesis. They focused on red howler monkeys in the jungles of French Guiana. As if to demonstrate its own evolutionary potency, three-color vision emerged again, independently, in howler monkeys in the Americas about 13 million years ago. It's guesswork to say what made it so successful, but there is one obvious candidate: color vision helped primates spot ripe fruit.

Howler monkeys favor the fruit of the *Chrysophyllum lucentifolium* tree, which have tough skins that they rip open with their teeth, and large seeds that pass through the primate digestive system. The fruits ripen to a rich blend of yellow and orange, an ideal contrast with surrounding greenery. For days a team of researchers camped out in the lowland forest, beneath a leafy canopy about a hundred feet above them. They followed groups of monkeys as they scrambled through the treetops, collecting the remains of devoured fruits.

Using a spectrometer to measure the wavelengths in the colors of the plants, the scientists found that the pigments in the retina of the howler monkey are almost perfectly attuned to the task of spotting ripe, yellow fruit amid the foliage. This was apparently no accident; the *Chrysophyllum* fruit's colors occupy a very narrow band of the spectrum. Natural selection seemed to have finely tuned one to the other, producing advantages for both: food for monkeys and a way for the fruit trees to disperse their seeds. (Other foods may have also played a role: in some primates, three-color vision may have evolved to spy nutritious young red leaves among green foliage when fruits were scarce.)

Colorful fruit, then, wasn't just a rare, tasty treat or even an important element of the prehistoric food pyramid. It was part of a broader survival strategy. The nocturnal cycles of the monkeys' ancestors now gave way to daytime sweeps. High in the trees in the light of day, colors supplanted scents. Smell, so central in the development of intelligence and awareness, receded. Now vision took point. This tilt from one sense to the other is written into our genes: primates with three-color vision have fewer working olfactory receptor genes than those without it, meaning they can detect fewer scents.

Forests and jungles are filled with edible leaves, but fruit trees are scattered, and some bear fruit only at certain times of the year. Survival depends on some level of planning. To keep eating, an animal has to remember where the best trees are and when they're likely to produce edible fruit. Fruit is a true prize, and it takes smarts to get it. Fruit-eating chimpanzees, bats, and parrots have bigger brains relative to body size than leaf-eating gorillas, grub-eating bats, and most other birds, respectively.

Unlike solitary *Morganucodon*, ancient monkeys moved and worked together as a group, communicating by sound, glance, or gesture. Superior eyesight helped here, too. Their eyes were set forward in the head, giving them three-dimensional vision—oddly, this anatomy is typical of predators, not scavengers. It puts potential prey in the center of the visual field, where it can be swiftly identified, evaluated, and attacked. But for primates, depth perception made it easier to spot the movements of stealthy, camouflaged predators, and to travel swiftly through networks of branches in low light, where one wrong move could mean a deadly fall. With just one pair of eyes, focused forward, each individual's survival depended on the whole group acting as a unit, with multiple pairs of eyes pointing in all directions.

The demands of the hunt would have also favored ever-more-expressive faces. The brains of apes and humans have much larger visual cortices than those of other mammals relative to body size, and bigger nerve centers for making faces. Blunt expressions of fear, disgust, and pleasure present in all mammals broke loose of their roots as involuntary reflexes and added layers of individual subtlety. A glance could convey volumes. Like a band of marines, the monkeys functioned as a food-gathering unit, their feasts anticipating the present-day communal meal.

Seared Fish with Olive Garnish; Fricasseed Gazelle

In a system of basalt caves near the edge of a lake, early humans built a hearth circled by stones. Their community lived amid plenty: the lake had schools of catfish, tilapia, and carp. Crabs scuttled over the sand. Turtles meandered. On nearby hillsides, wild olives and grapes were there for the taking. The women and children would gather food and toss it into the fire. They'd watch it singe and crack, then push it out with sticks, popping the best bits into their mouths so their tongues burned, savoring the carbon-flecked flesh of fish and fruit. Men sometimes tracked and killed animals for meat, but more often they found leftovers, the remains of a deer or an elephant freshly killed by some predator. They stripped the meat and roasted it, blood and fat sizzling.

Starting about a million years ago, groups of some close relative of *Homo sapiens* occupied this campsite, located at the Gesher Benot Ya'aqov cave in the Hula Valley in modern Israel. It was a pleasant spot, rimmed by mountains that cooled the desert climate. Fresh water bubbled up from mountain springs and flowed into a river just to the south. The groups stayed for tens of thousands of years, until a mudslide or cave-in entombed the campsite roughly 780,000 years ago. In 1935, archaeologists from the Hebrew University of Jerusalem discovered the cave and began a meticulous, decades-long excavation. They uncovered an amazing story of prehistoric dining, and a snapshot of how flavor emerged from its animal origins.

Excavators uncovered clusters of burned flint shards, as well as chunks and splinters of singed ash, oak, and olive tree branches. Studying these in the 1990s, archaeologist Naama Goren-Inbar deduced their condition could not have been the result of a random wildfire. Fires caused by lightning strikes burn briefly across wide areas, and at lower temperatures than man-made fires, which are carefully tended to focus their heat. The food items had been roasted at high temperature. The

Gesher Benot Ya'aqov cave dwellers had achieved the Promethean ideal: they could control fire.

They were using it to cook. Husks of burned grains and acorn shells were also found in the main hearth area. The residents had roasted seeds from prickly water lily plants, water chestnuts, olives, wild grapes, and holy thistleberries. There were cooked fish bones and crab claws, as well as bone fragments from deer, elephants, and other animals. Fire was only the most potent of a whole suite of tools used in food preparation: these early humans had a kitchen. One area was devoted to gutting fish. A space used for processing nuts had hammerstones and pitted anvils that had been used as bases for smashing the shells of acorns before roasting them. More anvils, these used for making flint tools, were nearby.

Their own remains have not been found—they may have disintegrated over a million years, or been buried elsewhere—so it's not clear what, exactly, these early humans were. They may have been members of *Homo erectus*, a species whose brain was about three quarters the size of that in modern humans, and who had a facility for toolmaking. *Homo erectus* had migrated out of Africa by this time, ranging as far as the Caucasus and into East Asia before disappearing about three hundred thousand years ago. Or they could have been another, perhaps still-unknown, predecessor to modern humans. Either way, they were radically different from their immediate ancestors.

"They were pretty impressive. One may say that they were pretty modern," said Goren-Inbar. "They knew the cycle of life of many animals and their drinking, eating, and social habits. They knew what plants to eat, and they knew where to go and grab raw materials to produce the stone tools: basalt, limestone, and flint. Those materials are very different, and they had to go and pick them up in different places. Even the fracture mechanics are very different, so making tools out of each required different skills. All in all, they were very sophisticated."

Over a few million years, a blip in the history of life, toolmaking, talking, self-aware beings evolved from groups of apes living in trees. The Gesher Benot Ya'aqov site offers a tantalizing glimpse of this transformation, in which taste, smell, sight, sound, and touch coalesced into our own flavor sense—a new type of perception that helped give birth to the human form and to culture.

Human evolution bears some resemblance to what happened in the Cambrian explosion and many times between: in the never-ending search for the next meal, bodies grew more agile, perceptions clearer, brains larger, behavior more complex—and flavors richer. But each story is different, each species's sense of flavor the result of a singular set of evolutionary conditions. As our monkey ancestors munched on fruit, natural selection pushed the tastes of other mammals in radically different directions. Whales and dolphins, which evolved on land, lost the ability to taste sweet, bitter, sour, and umami when they moved back into the sea, leaving only a sensitivity to salt—perhaps because most swallow fish whole and have no need to taste them. On their diet of meat, cat species grew insensitive to sweetness. And after they abandoned meat for bamboo, the ancestors of giant pandas could no longer taste savory umami. Humanity's emergence was a singular event, the result of an unlikely series of twists. If geography, habitat, natural selection, and plain luck hadn't converged in exactly the right way, we wouldn't be here.

Exactly how this happened is mysterious, but there are clues in the archaeological record and hints in our own anatomy and behavior. One element was nearly constant chaos.

Early humans lived on an ecological precipice that was always giving way beneath their feet. Around the time monkeys developed three-color vision 23 million years ago, the African continent shuddered and split. The ground over the fault collapsed, and rising plateaus on either side blocked the passage of rain clouds. That and other climatic changes dried out the African jungles and fragmented them like a dropped jigsaw puzzle. The forest scavenger's mix of fruits, nuts, leaves, and insects that had sustained monkeys and apes was scattered farther and farther apart, separated by dangerous open spaces. Natural selection went into overdrive; dozens of ancestral human species branched off in these changing environments.

Roughly two million years ago, the ground suddenly opened up under the feet of an adolescent male and an older female. (It's unknown whether they were together, or whether this happened to each separately.) Each fell dozens of feet into a vaulted underground chamber. They landed hard on the bones and rotting carcasses of other animals, and either died instantly or lay gravely injured until their end. Over time, layers of grainy, cement-like mud encased and preserved their remains.

In 2008, nine-year-old Matthew Berger was chasing his dog near an archaeological dig in the dolomite hills outside Johannesburg, South Africa, when he tripped on a log. "Dad, I found a fossil!" he shouted to his father, paleoanthropologist Lee Berger. It was the remains of the adolescent boy, who would have been four feet two inches tall. The elder Berger soon found bones from a female skeleton. They were the first of their species ever found, dating to just under two million years old and dubbed *Australopithecus sediba*. (*Sediba* means "fountain" or "wellspring" in the local Sesotho language.) Since then, the remains of an adult male and three infants have also been unearthed at the cave site, known as Malapa.

Australopithecines were descended from the first human ancestors to split off from the ape family tree, a few million years before the time of the Malapa fossils. "Lucy," from the related species *Australopithecus afarensis*, is the most famous such fossil; her 3.18-million-year-old bones were discovered in Ethiopia in 1974. She walked upright, but had long arms and powerful hands for grasping branches. The *sediba* pair lived a million years later than Lucy. They had the larger brains and nimbler bodies typical of later species. But when it came to food, they were curiously backward, standing at the threshold of change, but seemingly unable to cross it. The fossils were unusually complete and revealing, given their age, and among the remains were jaw fragments with nearly perfectly preserved teeth. As any watcher of police procedurals knows, dental records tell a story: what their owners ate, how they ate it, who they were.

To reproduce that two-million-year-old menu, scientists led by Amanda Henry, a paleobiologist at the Max Planck Institute for Evolutionary Anthropology in Leipzig, Germany, analyzed residues on the teeth. The plaque contained telltale signatures of various foods, microscopic flecks of plant matter called phytoliths. (As the name, Greek for "plant-stone," implies, phytoliths are composed of silica from the soil that plants absorb and distribute through their cells. When the plant matter rots away, the silica remains, providing an identifiable afterimage of the cells.)

Henry expected that the pair had lived on a savanna diet, heavy on grasses and roots, consistent with the environment in which they lived. But when she analyzed some of the tartar from their teeth, she was surprised. The *Australopithecus sediba* diet came almost entirely from the shrinking forests, which contained a different carbon isotope than savanna roughage: nuts with hard shells, broad leaves pulled from bushes and reedy trees growing low under the forest canopy, bark stripped off the younger trees and chewed like prehistoric jerky. Sometimes they would have eaten fruit, but such finds would have been rare. Bitter, leafy, herbal flavors were the highlights of eating.

It was a taste mystery. They could strike out across the savanna anytime it suited them. In order to keep eating their forest-based diet, they would have had to travel far, moving across grasslands, ignoring the food they offered. This diet was, on some level, a choice. Perhaps the flavors and textures of savanna foods repelled them. Did other groups behave differently? Did this group later change its behavior, or die when its favored foods ran out? It's sad to think of this species employing its emerging intelligence on a quest to keep eating a familiar but increasingly spartan diet, seemingly ignoring one key to its survival.

Shifting habitats forced human evolution onto an improbable path. Food sources grew unreliable and farther apart, so bodies became more upright, lean, and mobile. Brains grew larger to devise more sophisticated strategies for getting food. But these two trends conflict.

Compared to that of our closest relatives, chimpanzees, the human body is an improbably fragile vessel. Chimps have big guts and large, powerful jaws, and can open their mouths twice as wide as we can. Smaller human jaws and faces are traced to a 2.4-million-year-old mutation in a gene that makes myosin, a muscle protein, and produces weaker, finer muscles. The human gut is also small. Yet our brains are large, and demanding. An adult human brain consumes about a quarter of the body's energy. In other primates, it's only a tenth. On paper, this anatomy looks like a recipe for disaster. Chimps must spend many hours of each day chewing to sustain themselves. How did our ancestors ever eat enough to survive?

The *Homo sapiens* body works for one principal reason: bigger brains helped humans create better, tastier food. Our ancestors made up for their physical weaknesses by becoming skilled hunters and chefs.

In the 1930s, the legendary anthropologists Louis and Mary Leakey excavated a trove of fossils in Kenya's Olduvai Gorge that illustrated this progression over two million years. The earliest tools, from the time of the australopithecines and before, had been fished from the smooth quartz and basalt of the Omo River cobble and smashed and cracked to produce a flat surface that could be used for pounding. As time went on and more sophisticated species appeared, a craft developed: rocks were chipped away to create a characteristic spade-like shape of concave impressions and edges. Such a tool could be used for chopping and scraping. The most obvious use for such implements was butchering animals, and excavators also found stone tools and animal bones marked by cuts and hammer blows.

Meat became a dietary staple for members of our own genus, *Homo*. This changed eating forever. Unlike industrially produced meats, which are succulent and fatty, wild game is exceedingly tough. Cutting and tenderizing made it possible to eat more game. Starchy roots, another important staple, could be sliced or mashed. In other words, food was partially digested before the first bite was taken. Now constant chewing was no longer necessary, and meals were briefer and filled with strong tastes from start to finish: the savory, umami flavors of raw meat, the iron bitterness of blood, the richness of fat, the odd complexities of brains and kidneys.

Then came fire. It might have started like this: A lightning strike ignited the savanna scrub, the breeze sending a wall of flame dancing through the grass. Animals panicked and fled in all directions, eyes mad with fear. But a few dozen pairs of more practiced, nearly human eyes assayed the scene from a distance. They had seen this many times. They gauged the wind and the direction the flames were advancing, and moved together to make way, reaching a slight rise in the land for a better view. They would have felt the heat on their faces and chests as the blaze passed by, and a rush of excitement. After waiting for things to cool off, they inspected the charred wake of the fire, scanning the ground and the bushes for food. Heat-scarred fig tree branches and nuts littered the ground, their shells cracked by the heat. Perhaps one of the group swept up a few nuts and tasted one. The flesh had turned tender, and the flavor, the richness of seared fats under the charcoal, was delicious. Nearby, others also ate cooked figs, hot juice running down their cheeks.

The above description is based on primatologist Jill Pruetz's observations of savanna chimps, who maneuver around wildfires, then move in afterward seeking treats. Australopithecines and their descendants likely employed similar strategies, developing a feel for how to manipulate flame. Chimps, in fact, appear just a conceptual step or two away from controlling fire and cooking. Kanzi, a bonobo (a species of chimp) at the Iowa Primate Learning Sanctuary in Des Moines, became fascinated with fire at a young age. He repeatedly watched the movie *Quest for Fire*, about early humans struggling to rekindle their hearth, mimicking the actors and building small piles of sticks. When his keepers taught him how to light a match, he began setting fires. He'd manage them, tossing on extra wood when the flame started to die. Soon he was cooking: he'd take a marshmallow and put it on the end of a stick, and later began using a frying pan to cook hamburgers.

Like our ancestors, bonobos know that cooked food tastes better. Roasting makes meat tender, the toughest tubers mushy, and eggs palatable. Intense heat triggers a series of distinctive chemical reactions that allow flavor to bloom. At around 300 degrees Fahrenheit, the tightly coiled proteins in the muscle fibers of meats begin to break up and unwind. Their uniform shape is replaced by thousands of different configurations, which then clump together in a process called denaturing. The meat turns tender. Then amino acids combine with sugars, the start of a chain reaction that spins out thousands of distinct, flavorful chemicals in trace amounts. This process is known as the Maillard reaction, after the French physician and chemist Louis Camille Maillard, who discovered it a century ago. The Maillard reaction also generates pigments, turning baked bread, cooked meat, and roasted coffee beans brown. Today, manipulating the Maillard reaction is a cornerstone of food science.

The Gesher Benot Ya'aqov cave site's million-year-old hearths are the earliest widely accepted evidence of cooking, and archaeologists have discovered many more suspected ancient hearths dating back to four hundred thousand years ago, the time of the immediate forerunners of modern *Homo sapiens*. But there is evidence that cooking transformed human biology, and with it, the human flavor sense, sometime between one and two million years ago, providing the crucial calorie boost larger brains demanded.

Richard Wrangham, a Harvard primatologist, looked at the mechanics of eating and digesting raw food and wondered if it could really provide enough fuel for *Homo erectus* to survive. After the calories burned in chewing and digesting are taken into account, raw meat almost isn't worth the time and energy expended to consume it. Karina Fonseca-Azevedo and Suzana Herculano-Houzel of the Federal University of Rio de Janeiro calculated exactly how long such a raw meal would have lasted. Using data on body and brain size of primates, along with information on the time each species spent eating, they projected that a *Homo erectus* eating raw food would have to chew for eight hours—leaving him little time to get the food, and none to do anything else.

Cooking solves this problem by making food easy to eat and to digest. There's time to obtain, prepare, and savor a meal. And when food can be consumed in small, concentrated bursts, the unlikely combination of a small gut and a big brain starts to make sense. "Humans are biologically adapted to eat cooked food," Wrangham said. He did a number of experiments to test this idea: in one, he and Stephen Secor, a University of Alabama biologist, fed pythons diets of raw and cooked meat, and found they expended far less energy digesting the latter. Wrangham concluded that cooking must have been instrumental in *Homo erectus*'s large burst of brain growth starting around two million years ago.

Given the limited archaeological evidence of cooking fires more than one million years old, this theory is controversial. (Wrangham points out that evidence of fire use tends to disappear over time.) It also doesn't account for a second burst of brain growth after one million years ago, leading up to *Homo sapiens*, that has convinced many anthropologists that early humans began to cook later. But if the theory is true, a cooked diet had a large hand in our evolutionary success and anatomy.

As brains grew, natural selection redesigned the entire human head, including the interior of the mouth and nasal cavity. Smell returned in a new guise. In most mammals, a bone called the lamina transverse divides the nasal cavity. Chewing food liberates aromas in the back of the mouth, but this bone keeps them from reaching the nose, allowing animals to focus on smells around them. As apes evolved, the lamina transverse disappeared. Then, in humans, the passage from the mouth up into the nasal cavity shrank. It was merely a few centimeters' difference, but it supercharged our ancestors' capacity to experience flavor. As people chewed, a cascade of aromas reached olfactory receptors via this back passage.

Smells had tightly knotted our ancient ancestors' expanding awareness to the world around them. This anatomical legacy is still with us. As it was in the earliest mammals, the human olfactory bulb remains just a single synapse removed from the neocortex, where sensations become perceptions. This isn't true of the other senses; taste signals pass through the brain stem and hypothalamus before they reach the neocortex. Smells are unfiltered, immediate. As they entwined themselves with taste and the other senses during meals, flavor came alive.

• • •

At the Gesher Benot Ya'aqov site, people likely gathered for meals, savoring cooked fish and deer meat dripping with bubbling fat and the crackle of seared skin. They ate, drank, talked, and rested, satisfied. They'd reached the last link in a long chain of cooperation—planning, gathering, hunting, butchering, preparation—and the reward, a feast and fellowship.

In his second book on evolution, *The Descent of Man*, Darwin linked the rapid expansion of human intelligence to man's social nature: our talent for communicating, and for living and working together as a unit. The hardships our human ancestors faced likely pulled them together into tight-knit groups. A group of chimps in southeastern Senegal that Jill Pruetz studies follows this dynamic. Most chimps live in woodlands. But this area is mostly savanna, and food is sometimes sparse—conditions that have forced the Fongoli chimps, nicknamed for a stream in their habitat, to work more cooperatively. They form a larger, more cohesive group than typical woodland chimps, and are more willing to share food; in one encounter Pruetz observed, dominant males declined to challenge a hungry female who wanted to take fruit from a pile they'd made. They also fashion basic tools: sticks to scoop termites out of mounds, and spears to skewer tiny creatures known as bush babies that slumber in the nooks of tree branches. This yields a few ounces of meat.

One might expect to find that animals belonging to larger groups, with more complex dynamics, had larger brains. In the 1990s, the California Institute of Technology's John Allman set out to investigate this theory among primates. He was surprised to find that primates with bigger brains relative to body size *didn't* form larger social groups. But when Robin Dunbar of Oxford University narrowed the question down, he found something surprising. Overall brain size might not vary with group size, but the size of the neocortex did. Humans have the largest neocortex relative to body size of any animal; it's what gives the cathedral of flavor its magnificent architecture. It braids the basic urges and sensations around food together with thoughts, memories, feelings, and language. And it helps tie groups, and society, together.

Early humans had to collaborate to survive, developing complicated strategies to thwart adversity. Making tools and controlling fire require not only technical skill but knowledge that must be preserved and passed on to others. Hunting demands planning and teamwork. And as all backyard grill masters know, cooking meat depends on the skilled butchering of animal carcasses, fire management, and a dash of creativity. Over time, cooking became about more than just filling stomachs. Humans developed codes and customs around food. Using tools and knowledge to create flavor was the earliest spark of culture.

Every successful species adapts to its environment. Rick Potts, a paleoanthropologist who directs the Smithsonian Institution's Human Origins Program, says that humanity's talent was more formidable still: our ancestors adapted not just to different environments but to the hard reality that those environments are always changing.

This is one explanation for the great diversity of flavors and cuisines in the world today, and for a certain plasticity in human flavor sense that other animals lack: why we so easily develop a liking for things that are intrinsically unpleasant, such as bitter coffee or beer, or the heat of chili peppers or wasabi. The chaotic landscape of ancient Africa wasn't just savannas and scrub: it was dotted with volcanoes, rivers and lakes, plains and peaks, from more than 500 feet below sea level at Lake Assal in the Afar depression, the lowest point in

Africa, to 19,340 feet above sea level at Mount Kilimanjaro, the highest. Moving about these changing habitats was how humans first learned to live and thrive almost anywhere. Surviving the East African Rift's challenges was just the warm-up for the big show of world domination.

CHAPTER 3

The Bitter Gene

One day in March 1990, President George H. W. Bush banned broccoli from Air Force One. Broccoli belongs to the genus *Brassica*, the plant family that includes mustard, cabbage, and brussels sprouts, most of which have a similar defense: when cut, their cell walls break, triggering a chemical reaction that releases waves of alkaloids, complex molecules that the human body reacts to in many ways. The most obvious is their bitter taste.

When the news broke, nutritionists questioned whether this decision set a bad example for America's children. Incensed California farmers dispatched a cross-country truck caravan bearing ten tons of fresh-cut broccoli stalks to Washington. "I don't think the president was given broccoli when it was properly cooked," Julia Child weighed in. "Broccoli has to be peeled." At a state dinner, Bush was overheard jokingly complaining about the ruckus to the Polish prime minister. "The broccoli growers of America are up in arms against me," he said. "Just as Poland had a rebellion against totalitarianism, I am rebelling against broccoli."

Pressed for an explanation at a press conference, Bush made a now famous denunciation: "I do not like broccoli, and I haven't liked it since I was a little kid and my mother made me eat it. And I'm president of the United States, and I'm not going to eat any more broccoli!

"There are truckloads of broccoli at this very minute descending on Washington. My family is divided. For the broccoli vote out there: Barbara loves broccoli. She has tried to make me eat it. She eats it all the time herself."

"Cauliflower? Lima beans? Brussels sprouts?" shouted members of the press corps. Bush gave another thumbs-down to brussels sprouts.

George W. Bush shared his father's distaste. On his first trip abroad as president in 2001, he visited his Mexican counterpart Vicente Fox, a broccoli farmer. When his motorcade arrived at Fox's ranch in the low rolling hills of Guanajuato state, Bush got out and found himself standing against the backdrop of a vast field of broccoli stalks. The tangy scent of the cruciferous vegetable enveloped everything. Reporters asked him to comment. He hesitated for a second, then flashed a thumbs-down. "Make it cauliflower," he said.

Barbara Bush liked broccoli; her husband and son did not. Such stark differences are a basic feature of the sense of taste. Taste perceptions are genetic, programmed by DNA, traits passed down over millions of years that boosted the odds of survival in our evolutionary past. While both environment and life experience play a role in taste and flavor, the variety in human DNA is one of the main reasons why, like snowflakes, no two flavor senses are the same.

The great range in human taste perception makes it unique among the senses. The sensitivities of vision, hearing, touch, and smell vary only modestly from person to person. To survive, after all, our ancestors needed to live in more or less the same sensory world. Fragile, warm-blooded bodies function only within certain thresholds of heat and cold, so humans have similar tolerances for those. The rods and cones of our retinas tend to detect the same color wavelengths and play of light and shadow. The cochlea, the snail-shell-shaped organ in the inner ear, picks up common levels of noise and a range of pitch. And the olfactory epithelium in our noses discerns a similar array of incoming smells.

But the sense of taste is a sentinel, chemically testing everything that enters the mouth, so it has been molded by everything our ancestors ate and drank over the eons. It never occupied a single sensory world, but many. This is especially true of the taste we call bitter.

Bitterness originated as a biological warning system to keep toxins out of the body. Jellyfish, fruit flies, and even bacteria can sense bitter compounds, indicating this basic aversion can be traced back to the dawn of multicellular life. Sea anemones, for instance, which first appeared 500 million years ago, can sense and vomit up bitter substances that enter their digestive tracts. More recently, this taste has evolved in animals in tandem with plants, which produce most of the world's bitter substances. Plants developed toxic defenses to kill infectious microbes and to protect themselves from being eaten. There are many thousands of plants, and bitter compounds are seemingly uncountable. Our taste for bitterness is a product of this diversity—and of the boldness of our ancestors, who, after departing Africa a hundred thousand years ago, lived in and sampled the plant life of every habitat on earth.

A bitter substance on the tongue triggers an electrochemical cascade in the brain, which produces distaste. The outward result is a distinctive frown: mouth turned down, nose scrunched, tongue jutting out, as if to expel the unwanted substance. Faces across the animal kingdom, from lemmings to lemurs, display variations of this grimace.

Yet humans have a love-hate relationship with bitterness that runs through all cuisine. The word "bitter" comes from the Indo-European root "bheid," meaning "to split," the same root as "bite." In the Bible, bitterness is a metaphor for the suffering of the Jews. The bitter herbs used in the Passover seder, *maror* in Hebrew—horseradish, and parsley or endive dipped in salt water—recall the pain of bondage in Egypt.

But bitterness tastes good (for those who tolerate it well) when combined with other flavors. If it disappeared, a spark would vanish from food. Broccoli and its relatives from the mustard family, including cauliflower, brussels sprouts, kale, and radishes, are the most cultivated vegetables on earth. In the South, collard greens are often braised with pork; the fat and rich flavors of the meat soften the bitter flavor of the greens, and the bitterness gives the smoothness a tang. Chocolatiers have spent the five hundred years since Hernán Cortés, the conqueror of the Aztec empire, brought cacao beans to Spain from Mexico tempering their natural bitterness with sugar and milk. An element of bitterness is essential to beer and pickles—and coffee.

To make coffee taste good, the ancient, implacable force of bitterness is first summoned, then brought to heel. To understand how this process works, I visited the headquarters of Gimme! Coffee, a small chain of cafés and roasteries based in Ithaca, New York. The Gimme! roastery sits in a converted farmhouse on the edge of town. Inside, Jacob Landrau was monitoring two gas-fired Probat drum roasters, vintage black contraptions made from hand-cast steel parts. Each consists of a steel drum rotating inside a frame, like a clothes dryer, heated with gas jets to temperatures between 200 and 400 degrees Fahrenheit over the course of a roast, which takes about ten minutes. During the summer, temperatures can top 100 degrees in the roasting room, which lacks climate controls—that's a luxury reserved for the coffee beans, which are stored in an adjacent room where heat and humidity levels are kept constant.

Raw, dried coffee beans are a pale green; they are seeds that have been removed from a reddish fruit, then soaked and cured. When chewed, they have a mealy consistency and a grassy taste that's not particularly bitter. Many substances contribute to coffee's bitter taste; the best-known is caffeine. But roasting itself is responsible for most of it, teasing out chlorogenic acid lactones, which break down to form phenylindanes as the beans turn dark brown in the final stages, making darker roasts more bitter.

Landrau uses a laptop to track the temperature inside the roaster, but his own senses also guide him. If there's too much heat, the beans desiccate; not enough, and they turn bitter too quickly. He follows the sound of the beans as they rattle around in the drum, their appearance, and their aroma, all of which change minute to minute. Each batch has its own character, based on the type and age of beans used, as well as subtle factors such as the atmospheric pressure, quirks in the roasters, and the time of day. All must be managed to trigger a particular set of chemical reactions that generate the perfect flavor. If there's not enough bitterness, the coffee is lifeless; too much, and it's undrinkable, like the day-old pot at the back of a 7-Eleven.

On the day I visited, Landrau walked me through a roast from start to finish. After nine minutes in the roaster, the shells of the beans began to break, making a popping sound against the drum. This is called the first crack. Sometimes there are two or three cracks, if you apply enough heat: that means more bitterness. Landrau turned the heat off for a moment before turning it back up. This began a new roasting phase, in which sugars break down into water, carbon dioxide, fatty acids, and an assortment of flavor compounds. The temperature peaked at 389.9 degrees. "If you continue to develop the sugars and just burn it, that's the secondary spot where you're going to get the bitterness again," he said. He must also watch for tipping, another warning sign for excessive bitterness, when black spots appear at either end of the beans.

Landrau shut off the flame and opened the drum. The beans, now a robust-looking medium brown, poured into a circular tray, where rotating blades pushed them around to allow them to cool evenly.

Later that day, Liz Clark, who trains baristas for Gimme! shops, drew a graph showing lines for three tastes—sour, sweet, and bitter—rising and falling over time. We were in the Gimme! lab, where new formulas and techniques are tested. The graph is an important guide for baristas, illustrating something called "the rule of thirds." Because different substances dissolve at different rates, a single shot of espresso contains many flavors that emerge at different times as purified water passes through ground, roasted beans. Baristas must gauge the fineness of the grind, the water pressure, and the changing form of the drip as it emerges from the bottom of the filter and falls into the cup.

Clark asked a barista to tamp some finely ground espresso powder into a filter basket. It was a blend called Leftist ("Rich chocolate, caramel apple. Baking spice finish," its ad copy said). The barista inserted the basket into the espresso machine and gingerly pulled the lever, forcing 200-degree water through the grounds at nine times atmospheric pressure. As the shot dripped out, she divided it into three cups.

The first was dark brown, syrupy, and intensely sour. The

second was thinner and reddish, with a slight sweetness. The last was a pale, sandy color; "blonding" is a sign the shot has reached its endpoint. It was bitter. Individually, each cup tasted terrible, the bitter one most of all. Yet when the three were combined, the flavors played off each other delicately. This process can easily go awry: espresso machines are very sensitive, the flavors they produce temperamental. "You can really get to know someone's personality by the way they pull the shot," Clark said. "Even though there is what could be seen as a very narrow set of parameters, there is an almost infinite amount of variation for finding delicious shots in there."

Decoding the exact meaning of this ancient signal, alive in our bodies and food, is one of the more vexing problems in human biology. It has challenged scientists ever since one day in 1930, when two chemists in a factory had a fateful argument.

They were tinkering with formulas for blue dye at the DuPont chemical company's Jackson Laboratory at Deepwater Point, New Jersey. Arthur L. Fox was pouring a container of a white powder, a substance called phenylthiocarbamide (PTC), into a bottle when he fumbled, sending a fine puff into the air. His colleague, Carl Noller, a visiting Stanford professor, was standing nearby and inhaled some. It traveled from his nose into the back of his mouth and onto his tongue. It tasted sharply bitter. Fox was surprised; he had also inhaled some powder, but tasted nothing.

Fox put a pinch of PTC on his tongue and assured Noller it was tasteless. Noller dipped his fingers into the powder and stuck them into his mouth and winced. They asked other lab workers to do the same. A spontaneous experiment unfolded, with the scientists and technicians acting as their own guinea pigs. The mysterious split in reactions was confirmed: some could taste it, others could not.

In 1930, scientists believed that people's tastes were essentially the same. When they differed, it was attributed to mood or temperament. A child's dislike of brussels sprouts was a matter of poor discipline, not biology. The PTC discovery shattered that conventional wisdom. "Tastes differ far more than anyone realizes," Fox told an interviewer. "Beets may actually be disagreeable to Mary, while Johnny loves them. Father may not be able to tolerate buttermilk, and Mother may find garlic revolting. These foods simply do not taste the same to them as they do to others."

Fox's accidental discovery had pulled back a veil over the inner workings of genes. In Fox's time, scientists knew that the human body had a blueprint made up of genes. But lacking knowledge of DNA, they had no idea what that plan looked like. Every feature of human biology had to be connected to genes in one way or another, but it was impossible to disentangle their influence from other biological forces such as environment, upbringing, and aging. Fox's discovery, a simple genetic trait easily identified by a taste test, was the stuff of scientific revolutions. It might reveal how genes evolved and how they responded to changing climates or habitats. It could expose unknown genetic differences between genders, cultures, and races.

The Augustine friar and botanist Gregor Mendel first stumbled upon such single-trait genes—and with them, the first basic understanding of genetics—in the mid-nineteenth century, when he tried to breed pea plants to produce violet-colored flowers. Working at the Abbey of Saint Thomas in

Brno, in what is now the Czech Republic, Mendel crossed white-bloomed peas with purple-bloomed peas. Instead of violet, only purple blossoms appeared. He then bred thousands of pea plants and studied the resulting colors. Purple blooms crossed with purple or white yielded purple flowers. Only when white was bred with white did white flowers grow. He never did get violet.

The colors, he guessed, were produced by basic units of heredity, one from each parent. Mendel called them "factors." Purple factors were dominant. This allowed him to statistically predict the prevalence of each color: of every four blooms, three would be purple, one white.

So-called Mendelian traits—characteristics based on variations in a single gene—revealed the actions of genes to the naked eye. They are rare in humans; in Fox's era, eye color and blood types were thought to be Mendelian, but they later turned out to be more complex. Now Fox had found a new one. He put the word out immediately, placing news of the discovery that some people were "taste-blind" for a substance, and others weren't, in *Science* magazine. He began a series of taste experiments. "It was established that this peculiarity was not connected with age, race or sex," he later wrote in a seminal scientific paper presented to the National Academy of Sciences. "Men, women, elderly persons, children, negroes, Chinese, Germans and Italians were all shown to have in their ranks both tasters and non-tasters."

In 1932, Fox had a voting machine installed onstage at the annual conference of the American Association for the Advancement of Science, the nation's premier scientific organization. He invited members of the audience to taste PTC powder and pull the lever for their preference; 65.5 percent were tasters, 28 percent non-tasters, and 6.5 percent detected other qualities. This showed that the gene or genes for PTC tasting—and a greater sensitivity to bitterness—were dominant, like those for Mendel's purple flowers. The genes for non-tasting, or insensitivity, were recessive, like those of the white flowers.

A scientific craze was born. Scientists fanned out across the world to conduct taste tests on people of different ages, races, and social standings. They carried vials of PTC powder at first, and later a more practical kit that employed paper dipped in PTC solution and dried; volunteers placed the slip of paper on their tongues.

These experiments didn't always go over well. Along the back roads of Depression-era America, rumors circulated that the tests were a eugenics project to sterilize impoverished men. One tester noticed that when he showed up at a farmhouse, the women would gather round while the men made themselves scarce.

The test had its pop culture moment in 1941, when a pair of investigators from the University of Toronto journeyed to a farmhouse in rural Corbeil, Ontario, to test some of the biggest celebrities of the era, the six-year-old Dionne quintuplets.

Born two months prematurely to a farming family, the Dionnes were the first set of identical quintuplets known to have survived infancy. At birth, each could fit in an adult's hand; together, they weighed only thirteen pounds. The quints ignited a global media sensation, becoming symbols of survival for a world suffering through economic calamity. This fame came at a price; Canadian authorities had taken custody of them at four months old, after their father, Oliva, had signed, then canceled, a contract to exhibit the girls at the 1933 World's Fair in Chicago. Authorities feared they might be exposed to germs, kidnapped, or worse. The government offered a marginally more congenial form of exploitation, imprisoning the quints in a kind of bubble. During their early years, a team of doctors and nurses cared for them at a specially constructed nursery equipped with galleries where tourists could view them. Millions of tourists filed past, casting shadows the girls could see moving eerily across one-way screens.

The Canadian government also made the quints medical subjects for scores of experiments, supervised by the psychology department at the University of Toronto. Their growth and development was obsessively monitored and analyzed. So for UT psychologists Norma Ford and Arnold Mason, testing the quints' tastes was the obvious, inevitable thing to do.

When the time came, Cécile, Yvonne, Émilie, Marie, and Annette were escorted into a room one by one to be tested. Their teacher, Gaetane Vezina, explained what would happen: they'd be given three-quarter-inch strips of paper to place on their tongues, which they would chew. Some would be plain paper, the control, while others would be laced with tastes, including salt, citric acid, saccharine, and bitter quinine.

Since the Dionnes had an identical genetic makeup, it was a surprise when their reactions varied and were subtly impressionistic. Cécile compared the salty paper to the host wafer used in the Catholic Mass, citric acid to cough medicine, saccharine to sweetened medicine, and, for some reason, quinine to maple syrup. These distinctions held until they tasted the PTC paper, where genetics unified their impressions into singular distaste:

Cécile: "Ce n'est pas bon!" (This is not good!)
Yvonne: "N'aime pas le goût!" (I don't like the taste!)

Émilie: "N'aime pas le goût, pas bon!" (I don't like the taste, not good!)

Marie: "N'aime pas le goût, pas du tout!" (I don't like the taste, not at all!)

Annette: "Oui, il est fort!" (Yes, that's strong!)

Taste tests for bitterness remain a staple of science. In recent years a chemical called 6-n-propylthiouracil, or PROP, has replaced PTC as the test substance of choice; it lacks PTC's faint odor of sulfur and possible health issues. During a visit to the Monell Chemical Senses Center, I took the test. Danielle Reed, a geneticist and taste researcher, poured a small amount of PROP solution into a paper cup. It was clear, colorless, and odorless. I sipped it. Nothing. Like Arthur Fox, and approximately a quarter of the US population, I was a non-taster. It made sense. I've liked beer, coffee, broccoli, and other bitter things as long as I've been an adult. Non-tasters tend to be insensitive to other flavors, too, one possible explanation for why I like spicy food, and have trouble telling fine wines apart.

Then I took a leap forward into the twenty-first century. My family and I spit into tiny plastic test tubes, sealed them up, and sent them to the genetic testing service 23andMe in Mountain View, California, named for the twenty-three pairs of human chromosomes. The company's genetic profiling technology traces your place in the human family: the continents your ancestors came from, your risk for possible diseases with genetic components, the amount of Neanderthal DNA you carry thanks to ancient inbreeding. The test also reveals which type of Arthur Fox's bitter gene you have. After a few weeks, I got the results from the company website. All of us were non-tasters. This meant both my wife and I had inherited two copies of a particular non-tasting variant of the gene from our parents, and then passed these on to our kids. (The tests also showed 3 percent of our genome was Neanderthal; about average.) This fit my son's profile, with his penchant for spicy foods. But it seemed to contradict my daughter's preference for bland ones.

Between Fox's time and ours, the human genome—all its genetic material—has been discovered, unspooled, recorded, and partially decoded. Person to person, our genetic code differs on the order of only a tenth of a percent. But that small amount accounts for vast differences in body type, skin color, disease risks—and taste.

In the 1930s no one knew what a taste gene looked like, how it worked, or how the tongue or the brain could distinguish bitter from sweet. There were tantalizing hints about what occurred in these strange domains, but they were nearly impossible to detect with the scientific tools of the time: too small for a microscope, yet larger and more complicated than the chemist's traditional bailiwick of molecular reactions in test tubes. One scientist called it "the world of neglected dimensions."

By the 1960s, Massachusetts Institute of Technology molecular biologist Martin Rodbell was able to describe the strange biology of taste and genes using the lingo of the then-dawning digital age. Cells, he suggested, respond to their surroundings like a computer handles inputs and outputs. Something called a receptor was in charge of input: it sensed certain things such as bitter molecules, or hormones. Like flipping a switch, this triggered an electrical reaction inside the cell that beamed out a message across the nerves and to the brain, or another part of the body. Rodbell called this switch the "transducer." Taste, in other words, could be understood as a simple form of computing. A braised steak, a cup of coffee, a bitter berry—all contain thousands of different substances. Taste receptors—each made by a taste gene—extract essential information out of the chemical chaos of lunch and turn it into a code that the brain can interpret, so it can then react.

The anatomy of taste is a testament to just how wrong the original tongue map was. The average human tongue contains about ten thousand taste buds—tiny structures found on the visible, nub-like papillae. During a meal, the mix of food and drink in the mouth enters a bud via a single, pore-like opening at its tip. A bud is a knotted clump of fifty to eighty specialized cells, each detecting one of the basic tastes. One part of a coiled receptor protein protrudes out of a cell, the other part sits inside. The outside strand grabs molecules floating by, forming a temporary chemical bond. This makes the loops inside the cell pull apart, like the stems at the bottom of a bouquet when the middle is grasped too tight. This signal, essentially flipping the nerve cell to "on," triggers the cascade of signaling from the tongue to the brain that culminates, a tenth of a second later, in perception, awareness: *Ah, sweet. Ugh, bitter.*

The gene responsible for creating a bitter receptor was discovered, perhaps appropriately, not in a lab but in a computer database. By 1999, the first taste genes and receptors for sweetness, and closely related umami, had recently been isolated, and the scientists responsible had turned their attention to bitterness, performing experiments to isolate receptor cells. Meanwhile, they searched databases of the human genome, which had recently been decoded and published. A lot of it was still indecipherable strings of As, Cs, Gs, and Ts, the initials for DNA's amino acids. One day, Ken Mueller, a graduate student at a Columbia taste lab, was poring over that hash of letters when he noticed that some strings of code looked awfully like those for the genes of already known receptors: a bit like rhodopsin, which detects light, with a dash of pheromone receptor. It turned out to be code for a bitter receptor. They dubbed it T2R1. Within months, they found sixteen more. The current count is about twenty-three, give or take.

This mother lode explained a lot. There are only three genes for sweet taste, but the sweet receptor's task is simple: find sugars. Nature is filled with so many poisons that a whole repertoire of bitter receptors is needed to detect them. Over hundreds of millions of years, gene replications (like the kind that gave jawless fish and other creatures ever-more-powerful senses) doubled and quadrupled the array of bitter genes; natural selection tailored each of them to find different kinds of bitter. This is why only a few sugars taste sweet, while the number of bitter-tasting things is uncountable.

T2R1 (now called TAS2R38) turned out to be Arthur Fox's bitter gene. It's a strand of DNA found on chromosome number seven. Small variations in this sequence alter the receptor's chemical makeup and shape, creating the vast differences in people's ability to taste PTC and overall bitter sensitivity.

Isolating the DNA was, in some sense, the easy part. It was still not clear why nature programmed people with such diverse tastes—why some members of the Bush family loved broccoli while others despised it. There were deeper mysteries behind the bitter gene. Pursuing them inevitably led scientists back to where it all started: Africa.

At the height of Fox's taste-testing craze in the 1930s, a trio of English scientists became the first to explore the origins of the bitter gene. E. B. Ford, R. A. Fisher, and Julian Huxley were attending the annual International Congress of Genetics in Edinburgh in 1939 when they decided to do a PTC test on man's closest relative, the chimpanzee. They tracked down a Glasgow scientist with a supply of the powder, mixed it into varying concentrations, bottled it up, and set off for the Edinburgh Zoo.

They gave a medicine dropper full of bitter solution to one chimp. She spat it on Fisher. Another, enraged, tried to grab him. Bitter tasters, obviously. Six of the eight chimps were tasters and two were non-tasters, just how a random group of European humans might test. The zoo's orangutans, gorillas, and gibbons had a similar mix. This work was later cut short by World War II, but its implication was fascinating: at some point before humans and chimps diverged millions of years ago, natural selection sorted primordial ape populations into groups of bitter tasters and non-tasters. Whatever advantages these paired traits conferred must have been powerful, because both had persisted for so long.

It was a compelling theory. But once the bitter gene was decoded, it turned out to be dead wrong.

Biologist Stephen Wooding revisited the question in the early 2000s using modern genetic tools. It had recently become possible to track the course of evolution by studying an animal's genome. Over the eons, DNA mutates at a constant rate. Knowing this makes it possible to use differences in the DNA of related species to extrapolate how long ago they diverged from a common ancestor, or when a particular mutation happened. Wooding compared the human DNA code for Fox's bitter-taster gene in a human to that of a chimp. What he saw surprised him: the genes looked nothing alike. Somehow, different DNA codes produced identical taste experiences. Chimps and humans had each evolved the same traits independently, a finding that hinted they were even more potent survival tools than anyone thought.

Scientists have tracked these genetic signals through the past several million years and around the world, looking for what shaped them. Tasting and non-tasting had emerged first in chimpanzees, more than 5 million years ago, and more recently in early humans, between 1.5 million and 500,000 years ago—around the time that older species were starting to give way to the early *Homo sapiens*. Neanderthals, whose species split from a common ancestor with humans about 400,000 years ago, included both tasters and non-tasters, too. Carles Lalueza-Fox of the Evolutionary Biology Institute in Barcelona tested DNA from a 48,000-year-old fossil of a male Neanderthal unearthed, along with ten others, from a cave in El Sidrón, Spain. The Neanderthal turned out to be a taster.

When modern humans departed Africa around one hundred thousand years ago on their way to populating the world, they carried these genetic variations with them. Taste's small supporting role in this journey still echoes today in the taste for coffee and in produce choices in the supermarket, and shapes flavors throughout the world.

The crossing out of Africa likely occurred at what's now called the Bab el-Mandeb Strait near the southern outlet of the Red Sea, the narrowest stretch of water separating

North Africa from the Arabian Peninsula. Today, the strait is nearly twenty miles wide. But long ago, the Indian Ocean was hundreds of feet lower, and the strait was narrow and shallow, easy to cross by makeshift raft or maybe even by wading. Humans and their immediate ancestors were expert travelers. Similar crossings had occurred before, many times. Some early humans, like the people who tamed fire at the Gesher Benot Ya'aqov cave 700,000 years earlier, had made it as far as what is now Israel, about 1,400 miles to the north. *Homo erectus* had trekked to East Asia. And, as the humans camped on the beach, Neanderthals were hunting in Europe's forests.

These humans were more perceptive than their predecessors. Hundreds of thousands of years of evolution had made their brains larger, and their tastes simultaneously more robust and refined. The world had become one big lab for testing new tools and techniques for the hunt and the hearth. More than a hundred thousand years before the Bab el-Mandeb crossing, early humans started making earth ovens, digging deep pits and lining them with flat stones to focus heat for roasting game, roots, and vegetables. They may have smoked meat to preserve it, infusing it with new flavors in the process. They shared archaic recipes and passed them down the generations, tinkering as they went. These practiced rituals were a hedge against starvation wherever they traveled.

No one knows what drove them out of Africa for the unknown—possibly a famine, or some tribal dispute. Whatever it was, genetic evidence suggests that a small group of hundreds, or at most thousands, of people passed through this or several similar gathering points nearby, in a single exodus. Their descendants spread through the rest of the world, displacing (and sometimes interbreeding with) the only other human species around, Neanderthals. Virtually all non-African people in the world today are descended from the relatively small group that crossed at Bab el-Mandeb.

By chance, this founder group had quirky tastes. Genetic detective work indicates that in the Africa of this era, the human population had a fantastic range of sensitivity to bitterness; modern Africans still do. It makes sense that humans evolving in those diverse landscapes, encountering many different habitats and plants, would have a wide assortment of tastes. But the people setting out across the strait lacked this kaleidoscopic variety. Where it concerned Fox's gene, most were tasters, some were non-tasters, and only a few fell in between.

Theories vary as to the exact route early humans took after departing Africa. They may have hugged the shore, or headed northward and inland. At the time, the Arabian Peninsula wasn't the parched desert of today; it contained rivers, lakes, stands of trees, and stretches of savanna. Later, humans likely moved eastward across what is now Iran; then some moved northward, looping back toward Europe, while others pressed east into Asia. They colonized dense jungles and deserts, mountains and islands, from equator to poles, preparing and eating strange new foods wherever they went, eventually reaching the southern tip of South America about twelve thousand years ago.

A record of this seminal journey is now imprinted on our taste genes. The founder group's split between bitter tasters and non-tasters spread across the globe with their descendants, sorting much of humanity into one group or the other. Different habitats, climates, foods, and survival challenges somehow tuned the bitter division this way and that; some groups grew collectively more sensitive, some less. The evidence can be found in the record of Fox-style taste tests, more than a thousand studies done around the globe over eighty years. Plot the results, and they show that the split between tasters and non-tasters varies by geography.

In northeast Britain, nearly a third of the population can hardly taste bitterness, and among some ethnic groups in India, it's more than half. This may help account for the popularity of beer in Britain, and of bitter melon, the intensely piquant fruit, in Indian cuisine. Continue east, and far more people are bitter-sensitive: in some parts of China, as much as 95 percent of the population. This pattern holds for American Indians, most of whom are descended from Asian migrants. Travel to colder climates, and the balance shifts again. The Inuit of Greenland are among the least bitter-sensitive of early American peoples, perhaps because there's so little bitterness in their traditional fish-and-seal diet that they lost the capacity to sense it.

But another bitter gene had a very different fate. Nicole Soranzo, a researcher at University College London, studied a mutation in the bitter gene TAS2R16 that made people more sensitive to certain substances, including salicin, found in the bark of willow trees; arbutin, in bearberries; and amygdalin, in bitter almonds. This mutation is rare in Africa. Sampling the genes of populations around the world, Soranzo found that natural selection singled it out as a survival advantage elsewhere, and it spread like wildfire. Today upwards of 90 percent of non-Africans have it. Back in Africa, the trait never caught on.

The mysterious contradiction remained. Some forms of bitter sensitivity clearly enhance the odds of survival. But insensitivity does too. It must have some uses, or it would not be around today. What are they? In humans and apes, at least, bitterness serves as something more than a mere toxic substance alarm.

Alessia Ranciaro, an Italian biologist, was the intrepid leader of a small twenty-first-century taste-testing squad that sought answers. They traveled in Range Rovers from village to village through the bush of Kenya and Cameroon. At each stop, Ranciaro chatted with local elders. She asked for permission to administer taste tests and asked them to help persuade villagers to volunteer. Her team set up tasting stations far more elaborate than Arthur Fox's: each used a thirteen-bottle set of varying concentrations of several bitter compounds. Volunteers took a sip from each, rinsed, and spit. Ranciaro's team began with small concentrations and increased them gradually. "We go on this way until they start to say, 'I taste bitter,' or 'It tastes like lemon,' and they started to make funny faces," said Ranciaro. "When they taste bitter two bottles in a row, we can be confident. That is their final taste." They tested men and women from nineteen different ethnic groups and every traditional African lifestyle: hunter-gatherers, goat herders, farmers. They also drew blood for DNA tests, so that they could match sensitivities to genes.

Overseen by University of Pennsylvania scientist Sarah Tishkoff, this arduous study aimed high. By returning to Africa, the scientists hoped to finally identify the evolutionary forces behind our diverse bitter tastes. The answer would explain a lot about food, flavor, and human biology.

Tishkoff's team assumed the taste differences had to be tied to what people ate, and surveyed areas relatively untouched by modern society. "They're not eating McDonald's," said

Michael Campbell, who worked with Ranciaro and Tishkoff. For the most part, the people had been eating the same foods for thousands of years, perhaps as far back as the original exodus from Africa. People eating a meat-heavy diet might benefit less from an aversion to bitterness than a hunter-gatherer group harvesting pungent berries and roots. Over thousands of years, a meat-eating community might sift bitter-tasting genes out of the population.

When the results came in, however, the diet hypothesis—the theory of taste virtually all scientists subscribed to—turned out to be wrong. There was no evidence that food choices influenced genes—at least, not for the past five thousand years. Some older, deeper force seemed to be at work. This raised a provocative question: what if this ancient evolutionary signal was about something more than just taste?

It had been clear for a while that bitter sensitivity was part of a more complicated system in the body that reached beyond flavor. In the 1970s, Linda Bartoshuk, a taste scientist at Yale, noticed that many PTC tasters' sensitivity extended to sour, sweet, and salty tastes. They tended to avoid the powerful kick of chili peppers, wasabi, and ginger, too.

Bartoshuk set aside molecular biology and looked into the mouths of her volunteers. Many bitter tasters had a radically different anatomy from non-tasters, in that they had more fungiform papillae on their tongues. This meant they had more hardwired connections between the mouth and the brain: they perceived more intense taste sensations and more flavor information overall than other people. Some were ten thousand times more sensitive. She dubbed this group "supertasters." Their flavor experience differed from that of non-tasters. Their food was full of garish neon colors rather than gentle pastels.

Bartoshuk's finding could only mean that, in concert with many other genes, taste genes influenced the anatomy of the tongue and nervous system. Hundreds of studies show that the split between bitter tasters and non-tasters extends beyond food preferences: There are more women tasters than men. Alcoholics tend to be non-tasters. Tasting has been linked to diabetes, bad teeth, eye disease, schizophrenia, depression, gastrointestinal ulcers, and cancer. Some of these correlations may be random, but the preponderance of them indicates that bitter taste biology influences the whole body. Since the DNA of taste receptors was decoded over the last decade, it has been found all over the body: along the digestive tract, in the pancreas and liver, in the brain, and in the testicles. (Smell receptors have also been isolated in the liver, heart, kidneys, sperm, and skin.) "You can imagine a simple organism, a protozoan with a sort of mouth where the taste gene is expressed," said Roberto Barale, a biologist at the University of Pisa, in Italy. "When he starts to evolve and increase his size, the gene which controls expression of taste follows along. So the expression expands throughout the body, from mouth to stomach to intestine. And so on."

Receptors of all kinds make up a sprawling sensory web that wends through the cells in every living thing. Without them, life would be blind and inert. Certain yeasts use receptors to recognize the sugars they feast on, as well as pheromones, so they can mate. Fruit flies have them embedded on the outside of their bodies to sense changes in light and to trace the scent of ripe fruit wafting by. Plants have them too. Vertebrates have between one thousand and two thousand different types. There is no cell in the human body that doesn't contain them. They alert the body to changes in temperature and chemicals in the water or air. They are the switches for the body's internal communications, and thus the preferred target for most drugs, medicinal and illegal. They detect the presence of nerves firing, and rising or falling levels of hormones or neurotransmitters that move us to act on appetite, fear, or love. And they power sight, smell, and taste.

Still, the notion that "taste" would be at work all over the body was truly odd. Scientists at the Monell Center and the University of Pennsylvania tried to understand what use bitter receptors might be in the nose. They used cultures of human sinus cells with TAS2R38 receptors. To verify they had the right cells, they dosed them with PTC to see how they would respond. The receptors generated a faint electric current—the same signal they produce when they bond with a bitter molecule on the tongue. Bingo.

Then they gave the cells a sinus infection, flooding them with mucusy gunk. The bitter receptor alarms went off. The neurons sent an electrical message and released nitric oxide, a signaling molecule. Tiny cilia in the sinus cells waved faster and mucus production increased in response. That's how noses expel bad bacteria, or try to. Tasters, it seemed, were less likely to develop sinus infections than non-tasters. This would have been an advantage—unrelated to diet—for early humans colonizing colder climes.

There's also the gut. Like many small communities along the Mediterranean coast, the people of Calabria, the province that comprises the toe of Italy's boot, tend to be especially long-lived. Perhaps it's a consequence of the Mediterranean diet's emphasis on ingredients such as olive oil, fish, and red wine. But it might also have something to do with their sense of taste. Calabrian cuisine makes significant use of bitter vegetables, including eggplant, cauliflower, and spinach. The Bergamot orange, more bitter than grapefruit, is a local staple.

Roberto Barale was curious about the Calabrians' bitter tasting genes. As it happened, a group of Calabria residents was already taking part in a study tracking their health over the course of decades. Barale studied them, too, and found something provocative: a mutation in their TAS2R16 bitter gene was more prevalent in the elderly. And the older they got, the more likely they were to have it. Since people without it were dying earlier in life, something about this mutation seemed to contribute to longevity. Barale speculates this may be the result of bitter receptors in the intestines—although their function is currently unknown—somehow facilitating metabolism.

Until recently, most taste studies focused on the biology of the tongue and on perception. Now a new frontier has opened up. The bitter receptors dotting the body are part of a kind of shadow taste system. Unlike those on the tongue, they don't register in consciousness, and what they do is still mostly unclear. Flavor, in other words, is only the capstone of a vast, hidden system. It starts in the mouth with a burst of deliciousness, then disappears in darkness down into the gut, and from there its hand reaches everywhere in the body. It is an infinite mesh of sensors furiously sending and receiving messages as the whole body marinates in the chemical flux of the world.

The body's "taste" for bitterness may be responsible for the ubiquity of bitter foods today. Humans need bitter compounds; many are beneficial in low doses. Chewing willow bark is an ancient folk remedy for pain or fever—salicin is an anti-inflammatory compound related to aspirin. Bitter melon, a fruit grown in Asia, Africa, and the Caribbean, contains a suite of unpleasant-tasting chemicals that mildly lower blood sugar. As humans populated the globe, bitter-sensitive people may have helped their groups survive by detecting poisons; non-tasters might have tried more new things, bringing the others along when they found something with potential.

Humans can get used to anything. The Aymara people of Bolivia's Altiplano farm a very bitter type of potato, and their taste perceptions have adapted accordingly. Tests done in the 1980s showed the tribe was less sensitive to bitterness than Americans—yet every person tested was a PTC taster. Their sensitivity had dulled because their diets demanded it. Even at this new threshold, however, their tastes remained discerning: the point at which potatoes started to taste too bitter to them was exactly the point where they became toxic.

Like much else in the body, the flavor sense is a dialectic between genes and life experience. As people age, and sample ever-expanding varieties of foods, the networks of neurons in the brain responsible for aversive reactions shift. Bitterness mellows; for some it does a volte-face, becoming pleasurable. This capacity for contradiction, the strange yen to embrace the aversive, is what makes cuisine come to life.

It might have gone like this: Many generations after Bab el-Mandeb, a group of humans had migrated north of the Mediterranean, setting up camp in a valley. Culling the underbrush for food, they found sprigs atop twisted roots: plants from the genus *Brassica*, the wild ancestors of broccoli and mustard. What these lacked in tastiness, they made up for in nutrition: chemicals called isothiocyanates that stimulate the immune system and provide protection against cancer. To the eternal regret of President Bush and all modern broccoli-haters, this assured *Brassica*'s future.

CHAPTER 4

Flavor Cultures

A bsinthe, a green-hued alcoholic beverage made with anise and assorted herbs and extracts, has a reputation for mystery and danger. Invented in Switzerland in the 1700s as a medicinal elixir, it became the preferred drink of artists, writers, and bohemians, who were drawn to its sharp herbal flavors and the vivid highs it was said to produce. In fin de siècle Paris, it cast a spell. "What difference is there between a glass of absinthe and a sunset?" Oscar Wilde wrote. In *For Whom the Bell Tolls*, Ernest Hemingway described it as "opaque, bitter, tongue-numbing, brain-warming, stomach-warming, idea-changing liquid alchemy."

Among absinthe's ingredients is thujone, an intensely bitter chemical with a menthol fragrance obtained from the flowers of wormwood, a small shrub. (Wormwood extracts are still used as a traditional folk treatment for intestinal parasites, and to kill insects.) A century ago, high doses of thujone were thought to cause hallucinations and madness. Vincent van Gogh was a heavy absinthe drinker; in 1887, he painted *Still Life with Absinthe*, which depicts a tall, shimmering, pale green drink sitting next to a water decanter on a table in a Paris café. After he killed himself in 1890, the art world speculated absinthe had been responsible for everything from impairing his color perceptions—leading to his use of bright and off shades in paintings—to his mental deterioration, to his death itself.

Once known as the "green fairy," absinthe came to be called the "green witch" and the "queen of poisons." After a laborer went on an absinthe bender and shot his pregnant wife and two children in Switzerland in 1905, the Swiss government banned it. France followed suit in 1915; "absinthism" was blamed for eroding French culture. These fears lingered for decades. Even after the repeal of Prohibition in 1933, absinthe remained illegal in the United States until 2007.

Modern science has shown that absinthe got a bad rap. While thujone blocks the action of GABA receptors, one of the nervous system's principal signaling tools, it would take a massive dose to have any effect. A 2008 study of thirteen century-old absinthes found that each contained only trace amounts of thujone; overindulging in absinthe would cause alcohol poisoning long before thujone could do any damage. Scholars now believe van Gogh's decline was caused by some form of mental illness, combined with alcoholism.

With the legal strictures lifted, enterprising beverage makers set out to re-create and rehabilitate absinthe. One of them, Jedd Haas, built a distillery in a modest cinderblock room under an overpass in an industrial district of New Orleans in 2011. He and his partners named it Atelier Vie, French for "Life Workshop."

Distillation—the process of heating, evaporating, and condensing a liquid to purify it—dates to the ancient world (the term "distilled spirit" was coined by Arab alchemists, who thought that vapor contained the spirit of a substance). Distilled alcohol is a comparatively recent invention. Doctors in Salerno, Italy, began consistently making it in the twelfth century for medical use; in China, a form of distilled wine became a popular drink among the upper classes a century later. Making these drinks takes many steps. First, an alcoholic beverage is required; this may be wine (the basis for brandy) or a barley mash (whiskey) or fermented molasses (rum). It's heated in a still to a temperature above the boiling point of alcohol, 173.3 degrees Fahrenheit, but below water's 212 degrees. The alcohol evaporates faster, giving the vapor a higher concentration than the original drink. In the simplest kind of still, this vapor flows from the heated kettle into a separate container, where it is cooled and condensed into droplets, which collect in a third vessel. The resulting drink then may be aged and/or flavored.

Absinthe adds an extra twist to this process. First, herbs are infused into a previously distilled clear spirit. Atelier Vie uses rum; its hint of sweetness balances the bitterness of the herbs. This alcoholic "tea" is then distilled. The redistillation makes absinthe one of the most potent drinks on the market; Atelier Vie's is 136 proof, or 68 percent alcohol. (Scotch whisky is typically 40 to 50 percent alcohol.)

Serving absinthe involves a bit of chemical showmanship. First, Haas poured some of his absinthe into a glass. Instead of herbal green, this drink was a deep red, created by a secondary infusion of natural colors including hibiscus flower. He then placed an ornate, slotted silver spoon across the rim, set a sugar cube in the bowl of the spoon, and poured chilled water over it. As the sugar water and absinthe mixed, dissolved herbal compounds coalesced, turning the drink cloudy; this milkiness is called the *louche*, French for "shady."

A drink such as Atelier Vie's Toulouse Red owes its existence to centuries of refinements, technological advances, and cultural foment. It is many steps removed from the natural world, its original ingredients transformed beyond recognition. Its sensations and effects on the brain are like nothing found in any ancient hunter-gatherer's meal. These differences can be traced back to the birth of civilization itself about twelve thousand years ago, when both culture and the tools people used changed profoundly, and flavor with them.

At that time, the great post-Africa migrations were ending. The Ice Age was over, glaciers were receding, and a climate of warm, dry summers and cool, wet winters settled in across Europe and Asia. Wild grasses such as wheat, barley, and rye thrived, spreading over the Fertile Crescent, the area spanning the Tigris and Euphrates river valleys. People began eating these grasses, then cultivating them. In mountains not far away, others learned the tricks of herding goats and sheep. Cultivated crops and domesticated animals replaced the more diverse foods found in nature.

With this simplification of diet came a flood of food and flavor innovations. One of these rivaled the taming of fire: humans harnessed fermentation. Today, fermentation is the source of much of the flavor in the world, its signature found in a galaxy of consumables, which, in addition to spirits, includes wines, beers, cheeses, yogurts, tofu, soy sauce, and pickles.

A basic biological force, fermentation is the metabolic action of certain types of bacteria and fungi. These single-celled organisms belong to the microbiome, the sprawling populations of microbes that cover human skin, line our insides, and infest every square inch of the planet. One of the microbiome's most important jobs is decomposition: microbes feast on dead tissue, injecting its molecules back into the circle of life. Fermentation is a particular kind of decomposition, the breakdown of carbohydrates in the absence of oxygen. It has the felicitous result of making decomposing things taste better, rather than worse.

The by-products of fermentation include carbon dioxide, acids, alcohol, and a host of cast-off molecules. These were useless waste to the microbes, but they captured prehistoric imaginations. Their flavors were complex and provocative. Alcohol also altered brain chemistry, lowering inhibitions and smoothing social interactions. These new sensations shocked stunted palates, and altered the nature of flavor itself. Taste and smell are usually thought of as a series of chemical reactions in the mouth and nose. But flavor only comes to life at the other end of this system, in the brain, where chemistry is transmuted into sensation and consciousness. Just as the advent of cooking unleashed new flavors and nutrients that influenced the course of evolution, fermentation impressed itself on human biology and the mind.

There was no single "first" alcoholic beverage, cheese, or any particular fermented food. Like cooking, these items were probably invented a number of times, in more than one place. But they were profoundly different from cooked food. The tools of civilization gave prehistoric peoples a level of control over nature, specifically microbiology, that had never been achieved before.

All the elements for success were already in place in nature, waiting to be assembled. One of the most prolific, and useful, microorganisms on earth is a species of yeast called *Saccharomyces cerevisiae*. It is the hidden agent behind virtually all alcoholic beverages, as well as breads and other baked goods, which is why it's also known as baker's yeast. *Saccharomyces cer-*

evisiae is a supermicrobe. It can store lots of energy to survive lean periods, and it manufactures enough alcohol to kill off other yeasts, eliminating its competition. DNA from baker's yeasts still clumped to the legs of ant-like insects entombed in amber found in Poland and the Dominican Republic shows that it is tens of millions of years old.

There's an intriguing explanation for the ubiquity of baker's yeast: wasps. Wasps carry yeasts in their guts, and are attracted to fruits. In wine country, wasp nests often grow nearby as grapes ripen each season. Scientists at the University of Florence, Italy, looked for evidence of a connection. They caught wasps from colonies in Italy and analyzed their insides. Among 393 distinct kinds of yeast, the baker's variety stood out. The other yeasts waxed and waned during the course of the year, but baker's yeast was always present. It survived the cold by riding out winters in the guts of the fertilized queens. When young wasps departed their hives in the spring to form new colonies, baker's yeast went with them. In fact, the wasps are part of a global yeast transportation network; DNA evidence linked baker's yeast at Italian vineyards to many places in Italy and beyond: breweries, palm wine-makers, and bread ovens as far away as Africa.

This means that *Homo sapiens* was hardly the first species to encounter the products of fermentation. Nature has its own version of the alcoholic beverage, made by the action of baker's and other types of yeast on ripening fruits. The bud of *Eugeissona tristis*, the bertam palm, found in the West Malaysian rain forest, exudes a nectar containing as much alcohol as a craft-brewed ale. The ripening process marbles the tree's green fruits with bright colors and turns its pulp sweet; yeast ferments the sugars. Alcohol plumes are chemical Sherpas, carrying scents far and wide, attracting insects that assist with pollination, as well as shrews and slow lorises that spread the tree's seeds.

In Panama, howler monkeys habitually tempt fate by eating alcohol-laden palm fruit and drunkenly swinging through the trees. Biologist Robert Dudley tracked these monkey benders on a preserve on Barro Colorado Island in the Panama Canal. One monkey climbed up a thirty-foot palm and then leaped to another to grab the bright orange fruits clustered near its top. He sniffed each carefully. In twenty minutes, he had imbibed fruit infused with the alcohol content of two bottles of wine. And the more he ate, the more reckless his maneuverings through the branches became. Yet he didn't fall.

But it seemed to Dudley that the monkeys weren't just out to get hammered. They were discerning tastemakers, sampling different fruits to find just the right degree of ripeness, the tastiest balance between the sweetness of the sugars and the pungency of the alcohol, as if at a wine tasting. Drinking alcohol is something that primates have always done, Dudley suggested. He called this "the drunken monkey hypothesis": a certain amount of alcohol in the diet is normal, and shaped human brains and metabolisms. (Like the ills caused by eating and drinking too much sugar, alcoholism seems to be an unfortunate effect of civilization producing too much of something the human body evolved to tolerate only in limited amounts.)

Early humans left the rain forests to trek across savannas and through mountain passes, ultimately occupying many places that lacked ripening palm buds. But there were other opportunities to encounter alcohol. Patrick McGovern, an anthropologist at the University of Pennsylvania, believes early beverages developed out of a series of accidents. A beehive loosens and falls from a tree in a rainstorm. Yeasts swimming in the water and honey go to work, fermenting the mixture to mead in a matter of days. Honey + water + time is a simple recipe that humans would have noted, remembered, and shared. Gatherers might have harvested honeycombs, set them in hollowed rocks, and doused them with water, then left them in the sun.

At some point, people began to keep food in hollowed gourds, an invention that allowed a foodstuff's evolution from fresh to spoiled to be observed and tested. Human control of fermentation grew out of the ceaseless, and mostly losing, war on rot. When wild grapes were stored in gourds, some were crushed, breaking their skin and releasing sugary juice into the embrace of hungry yeasts. It frothed and bubbled. After a few days, it would become a pulpy, lightly alcoholic wine, tasty for only a short while before it turned to vinegar. The only choice was to drink and enjoy. Eventually, such accidents became recipes.

The earliest evidence of systematic beverage making was found in the 1990s at Jiahu, an excavated Chinese village along a branch of the Yellow River, settled about nine thousand years ago. An archaeological dig revealed evidence of surprising sophistication for a village dating back so soon after the dawn of civilization: A cemetery, holding hundreds of graves, sat beside clusters of mud huts. Social ranks had been established; some bodies were buried with jewelry, decorated tortoise shells, and ritual pottery vessels. People had fine toolmaking skills and, evidently, a talent for music: some graves contained flutes made from carved bone, the earliest ever found. They could still be played, and produced light, delicate notes. Archaeologists also uncovered some of the earliest known Chinese glyphs, the beginnings of written language.

They were etched into bones and shells: an eye, a window, and signs for the numbers one, two, eight, and ten.

Jiahu's artisans had also built earthen kilns to fire clay vessels, and many jars and pottery fragments were unearthed. When McGovern saw the clay jars for the first time, he was astounded: they were clearly beverage containers, resembling the wine amphorae of ancient Greece, though vastly older. Best of all, they weren't empty: a dried, reddish sheen lined the interiors of some, the remnants of a liquid.

Ancient drinks are McGovern's specialty. The work is challenging because alcohol doesn't leave traces: it evaporates quickly, and stray molecules are likely to be consumed by microbes. Most of his evidence is circumstantial, based on other ingredients he can decipher. Chemically analyzing the remains of the drink, McGovern found tartaric acid, which comes from fruit. The signature of beeswax pointed to honey as another ingredient. Finally, a test for carbon isotopes showed that rice had been present. Traces of tree resins (often used by ancient vintners as a preservative, and conferring a lemony tang) and herbs also appeared. The beverage would have been a cross between mead and wine, made from fermented honey, grapes, hawthorn berries, and rice. It was probably used in religious ceremonies. But it was also ordinary: drink-infused pottery shards were found both in graves and in homes. This was Jiahu's equivalent of a six-pack.

McGovern wasn't satisfied with knowing which chemical ingredients were used to make the Jiahu grog. He wanted to taste it. A sip, he felt, could help summon to life previously lost and inaccessible moments. It could help explain how civilization had changed humanity, reaching beyond the usual musty clues into an ancient people's lived experience—not just *what* they tasted but *how* they tasted, and felt.

In 1999, McGovern had teamed with Sam Calagione, founder of the Dogfish Head Craft Brewery in Milton, Delaware, to re-create an ancient brew from ingredients found in the 2,700-year-old tomb of King Midas, in Turkey. Calagione was motivated by naysayers who had denounced his early craft beers and their unexpected ingredients, such as juniper berries, chicory, and licorice root. "A lot of so-called purists, I'd say elitists, would say, 'You're screwing with the history of brewing!'" he said. Researching that history, Calagione found that modern beer recipes dated to a 1516 Bavarian law called the *Reinheitsgebot*, or "beer purity act," which mandated that the only ingredients of beer be water, barley, and hops (plus yeast, which wasn't known in the sixteenth century). The law is still in force in Germany, though imported beers are exempt.

Calagione set out to recover some lost pre–purity act brewing traditions. When he met McGovern, he said, "I could tell we were kindred spirits." Together, they tried to approximate ancient ingredients for the King Midas brew. To prepare a three-thousand-year-old Egyptian ale named Ta Henket, Calagione placed petri dishes laced with sugar at an Egyptian date farm to capture airborne yeasts, mapped their DNA to assure their provenance, and grew strains that were likely descendants of those used by the pharaohs. To make an ancient Peruvian corn ale, he spent four days chewing corn kernels so his spit would break down their starches into sugars.

For the Jiahu brew, McGovern and Calagione had only a list of likely ingredients from the original chemical analysis, with no quantities or instructions. The Jiahu people had gathered their ingredients from nearby forests and hillsides, as well as their own rice stores. Nine thousand years later, McGovern and Calagione would have to improvise. Chinese grapes that would have been a good match weren't readily available in the United States. Nor were the small, tart hawthorn fruits. The pair settled on canned muscat grapes, which are genetically similar to the wild Eurasian grapes the Jiahu artisans used. They were able to import hawthorn berry powder from China in fifty-pound bags.

The Jiahu people had grown and processed rice. This meant that neither brown rice (unprocessed) nor white rice (processed with modern technology) would do for the re-created brew. So McGovern and the Dogfish Head team used a kind of precooked rice with some bran and hulls still present. Finally, to set the stage for fermentation, the starches in the rice had to be broken down into sugars that yeasts could metabolize. For that, they turned to a concoction used in Asian cuisine known as *koji*, rice inoculated with a fungus, *Aspergillus oryzae*, that did the job. This was also a cheat, as villagers would have used more primitive means—spit, which contains the necessary enzymes. After a three-week brewing process, they had a beverage they called "Chateau Jiahu." McGovern found it delicious: effervescent, rich, brooding. (He also concluded it was an ideal complement to Chinese food.) I tried it myself. The honey gave it a smoothness, before a bitter aftertaste kicked in. It was not hard to imagine a summer evening, the sun going down, the rice and pigs tended to, a small fire burning, and perhaps the sound of a flute.

Throughout history, hungry microbes have fermented not only ripening fruits and honey but many edibles, launching a stream of culinary experiments. One began a few thousand years before Jiahu was founded, about four thousand miles away in a mountain range that spans present-day Turkey and

Iran, when a scene something like this must have occurred: A herdsman arose from the shadow of a lean-to. The morning light revealed a small herd of goats flecking the hillside below. He walked to a nearby pen made with woven branches for the female aurochs—a now-extinct wild ox—and her calf that he had captured. Aurochs were ornery, but this one was the calmest animal he had ever seen, and so with some effort he had managed to tame and breed her. He tugged on the aurochs's udder, filling a clay jar with milk. Milk made him sick to his stomach. But when he set it aside for a day or two, the lumpy curds—the simplest form of cheese—made a satisfying meal.

Aurochs were huge bundles of muscle, an irresistible food source, and herdsmen by this point had long experience handling flocks of goats and sheep. But both of those were docile creatures. Wild goats even sheltered in mountain caves; they were used to being penned. Aurochs were wild, mean, and unpredictable. Capturing and breeding them was nearly impossible.

Scientists led by Ruth Bollongino of the French National Centre for Scientific Research in Paris analyzed the DNA of modern cows and compared it with ancient DNA from fossils. They concluded that all cattle alive today are descended from about eighty wild animals, and that the original domestication probably happened in a single place in the Eurasian mountains—perhaps two—in an ambitious, or maybe just stubborn, project spanning generations. By comparing the DNA evidence with archaeological evidence of cattle herding, the scientists estimated that success took almost two thousand years. Dairy cheeses, the predominant type of cheese in the world, emerged during this arduous process, which changed human genes, biology, and tastes.

Before cattle were domesticated, almost all human adults lacked the ability to digest lactose, a sugar plentiful in the milk of all mammals. Children's bodies produced lactase, an enzyme that breaks down lactose, but they lost that ability as they matured. This is common among mammals after weaning: it eliminates the temptation to return to mother's milk. For the lactose-intolerant, drinking milk causes unpleasant side effects such as gas and diarrhea. Prehistoric herders would have found it maddening to have a huge nutritional resource at their fingertips that only children could drink.

But as milk ages, rod-shaped microbes start to break down its lactose. The Lactobacillales are an order of bacteria entwined with both the human body and the food we eat. *Lactobacillus acidophilus*, which is used to make yogurt, is found in the mouth and throat, small intestine, and vagina. Several species of *Streptococcus*, members of the same order, cause strep throat and pneumonia; others are used in making cheese. Spoiled milk is essentially partially digested, its lactose already broken down. This discovery was a nutritional bounty for our ancestors: adults who couldn't drink milk could tolerate dollops of bacteria-processed curds.

Cattle produce milk in vastly greater quantities than goats or sheep. As domestication efforts finally started paying off about ten thousand years ago—some generations after the lonely aurochs breeder's experiment—herdsmen began to swap their sheep and goats for cattle. Herding swept west and north across Europe, and with it genes enabling people to digest lactose. A positive feedback loop was under way: people able to consume milk and cheese fared better in dairying societies; dairying expanded as more people consumed milk and cheese. Today, only about 5 percent of northern Europeans are lactose intolerant. In parts of West Africa and Asia, where dairying never caught on, most people remain lactose intolerant.

A craft developed as dairyers made vessels to store milk and tools to handle it. Scientists have found the carbon signatures of milk fat on pottery shards from northwestern Anatolia, Turkey, dating from about 8,500 to 7,000 years ago, and clay strainers used to separate curds and whey in Kuyavia, Poland, about 1,500 miles away, dating from about the same period.

To temper their sour, lumpy cheeses, ancient herdsmen might have applied lemon juice or vinegar to hasten fermentation, or salt water, to brine the cheese. They could also have suspended a vessel of soured milk over a fire, letting heat condense the curds. Altering the balance of time, heat, and moisture in these processes made some cheeses sour, others mild and creamy, some dense and sharp. Some smelled like ripe sweat.

Prehistoric herders using animal stomachs as canteens likely found the milk they carried already congealed. The catalyst was rennet, a powerful, enzyme-rich coagulant found in the digestive systems of sheep and goats; it helps them digest milk by slowing its entry into the small intestine. Rennet attacks milk proteins called caseins, long, twisty molecules held together in loose, water-repellent globes. When broken, the globes begin to clump together. In cheeses, rennet creates a solid consistency. Cheese made with lactic acid bacteria alone tends to be mushy or crumbly, while rennet cheeses, including ubiquitous cheddar, Swiss, and gouda, are firm. It also provides a secondary dose of broken-down caseins for the *Lactobacillus* bacteria to metabolize and turn into flavor. Over the last two thousand years or so, still more microbes were added to the mix, most by accident at first. Molds grow on cheeses naturally; people found some tasty, and found ways to use them. One of them, *Penicillium roqueforti*, used to make blue cheeses, generates lipases, enzymes that break down fats as the cheese ripens, producing pungent tastes and blue-green marbling.

Recruiting these new microorganisms was, in essence, an ancient form of bioengineering. Unlike baker's yeast, some weren't naturally suited to the job; they had to be tamed. "We select for sheep that have lots of wool, cows that have lots of muscles, and plants that have larger fruits. If you start thinking about food, wine, cheese, and yogurt all involve the use of microbes that have been domesticated by humans. But whereas we know a lot about plants and animals, we don't know as much about what we did with the microbes," said biologist Antonis Rokas of Vanderbilt University.

Aspergillus oryzae, the *koji* fungus employed in making Chateau Jiahu, produces fine yellow-green filaments called hyphae, invisible to the naked eye, but sometimes coated with spores that give it the appearance of fuzz. It breaks down starches to sugars, which baker's yeast then converts to alcohol. Variations of this one-two combination are present today in all beverages made from grains, including beer, sake, and whiskey. *Aspergillus*-infused rice is used on soybeans to make soy sauce, miso, and other dishes. Japanese supermarkets sell it in sealed plastic bags. Rokas and his lab team compared the *Aspergillus oryzae* genome (sequenced by Japanese scientists in 2005) to that of its closest wild relative, *Aspergillus flavus*, as one might compare the genes, anatomy, and behavior of a cocker spaniel with those of a wolf to understand how dogs became dogs.

Aspergillus flavus is a scourge of agriculture and the source of a potent poison called aflatoxin, which causes liver cancer, acute hepatitis, and immune system damage. The two fungi share 99.5 percent of their genes; DNA evidence suggests *Aspergillus oryzae* probably originated from a single domestication of an *Aspergillus flavus* culture by some East Asian perhaps four thousand years ago.

Just as dogs were bred for friendliness and loyalty, the *koji* fungus was bred to make flavor. The first clue was its consistency. The *flavus* DNA varied a lot from batch to batch. Some of the *flavus* fungi were even nontoxic; that ancient brewing pioneer probably chose those (and made people ill if he chose others). But the *koji* fungi were genetically alike, all highly efficient at breaking down starches and producing flavorful by-products. One of their genes carried instructions for making glutaminase, an enzyme that helps produce the active ingredient in umami. A cluster of nine genes produced sesquiterpenes, compounds found in ginger, jasmine, and lemongrass that heighten aromatic sensations. Several of these molecules are manufactured and used by food companies today.

The torrent of new flavor molecules unleashed by fermentation galvanized ancient humans' senses of taste and smell. Flavor's great power derives from the synergies it creates between senses, different systems of the body and brain uniting to form something greater than the sum of their parts. Fermented foods, in particular, amplify this effect. "I am tempted to believe that smell and taste are in fact but a single sense, whose laboratory is in the mouth and whose chimney is the nose," gourmand and former French revolutionary Jean Anthelme Brillat-Savarin wrote in his 1825 book, *The Physiology of Taste: Or, Meditations on Transcendental Gastronomy*. A journey through the worlds of cooking and the senses, the book founded an enduring genre: the culinary essay. (Brillat-Savarin also penned the aphorism "you are what you eat.") In it, Brillat-Savarin suggested flavor was properly viewed not as a static phenomenon but as a process. As it unfolded, senses were activated—sometimes separately, sometimes together—before ultimately merging:

> He who eats a peach, for instance, is first of all agreeably struck by the perfume which it exhales; he puts a piece of it into his mouth, and enjoys a sensation of tart freshness which invites him to continue; but it is not until the instant of swallowing, when the mouthful passes under his nasal channel, that the full aroma is revealed to him; and this completes the sensation which a peach can cause. Finally, it is not until it has been swallowed that the man, considering what he has just experienced, will say to himself, "Now there is something really delicious!"

The mouth and nose aren't located far apart, but they are radically different in structure and function—it's surprising taste and smell are able to come together at all. Scientists have found just five basic tastes, programmed by a few dozen genes. Each is distinct, unvarying, and immediately identifiable in the complex mix of flavors in food and drink. Smells, on the other hand, are virtually infinite: there are as many as a million distinct ones, detectable by our four hundred kinds of olfactory receptors. The receptors bond with aromatic molecules in combinations far more complex than those involved in tastes. Smells are also subtler sensations. They blend seamlessly into flavors, submerging their identities into the whole. This combination of superior range and nuance makes smell the single most powerful component of flavor.

The human brain elegantly summons clarity out of the flux of smells and aromas. One day in 1974, Gordon Shepherd, a neurobiologist, went to a Maryland supermarket and bought a hunk of strong cheddar cheese. Shepherd intended to map how the brain interpreted aromas, which was then largely a mystery. The obstacle, as it had been since the time of the ancient Greeks, was analyzing a subjective experience. The activity of the brain's living neural networks was inaccessible. Ordinary X-ray machines could not capture neurons firing or blood flow in the brain. Implanted electrodes were sometimes used in animals and humans, but this was crude and imprecise.

Using a new method, a forerunner of today's fMRI scans, Shepherd and his colleagues at the National Institutes of Health injected rats and rabbits with a radioactive isotope that attached itself to areas of the brain where neurons were firing. As the animals sniffed the cheddar, elaborate patterns of activity were mapped in their olfactory bulbs. Unfortunately for the animals, the only way to see this was to examine their brains directly. After forty-five minutes of sniffing, the animals were euthanized, cross-sections of their olfactory bulbs exposed to X-rays, and the films studied with a microscope.

Each aroma produced a unique pattern that resembled an abstract pointillist painting. Smell, Shepherd concluded, was something like sight; each scent created its own distinct "image." In the eye, the retina converts light that strikes it into patterns of firing neurons in the brain that we experience as images. The olfactory bulb encodes scents and aromas in another type of pattern, which we experience as a smell, or as part of a flavor. The brain then refines these smell images further, adding contrast, creating crisp patterns as recognizable in their own way as the Washington Monument or the *Mona Lisa*.

Aromas, especially those in the complicated flavors of fermented foods, are instantly recognizable, yet elude description. Instead, they are typically described by analogy to something else: "a coffee smell," "smokiness." In this way they are a lot like faces. "We're very good at recognizing faces, but are very poor at describing them in words," Shepherd said. "Smells, too, are irregular patterns, ones we're not conscious of, yet the brain has to connect to the cognitive process underlying language to describe them. It's also difficult, after hearing them, to describe musical passages."

The waves of microbial activity in cheese leave behind a chemist's smorgasbord of alcohol, acids, aldehydes, esters, and sulfurous substances. Many launch into the air attached to evaporating water or alcohol molecules, forming evocative aromas. One substance found in Camembert, acetaldehyde, produces a pungent, nutty, yogurty flavor. A compound called 2-methylpropanal, found in gouda, has a malty banana flavor with a hint of chocolate, while butyric acid (gouda and cheddar) has a typical cheesy, sweaty, putrid aroma similar to sweat. Methional (cheddar) has scents evoking cooked potatoes, meat, and sulfur.

These aromatic images become delicate portraits of experience, etched into the nervous system as they're routed to parts of the brain involved in memory (the hippocampus) and decision making (the orbitofrontal cortex). In other words, smell literally links past and present. Because of its ancient roles mapping our surroundings and driving the evolution of the brain, smell is the only sense whose receptors connect so directly to these structures, only two synapses removed from the outside world. This gives it both immediacy and instant context, as the slightest whiff of a familiar scent can trigger a cascade of memories and feelings.

When Marcel Proust's narrator bites into a tea-soaked madeleine at the start of *In Search of Lost Time*, he is transported to his childhood village of Combray:

> But when from a long-distant past nothing subsists, after the people are dead, after the things are broken and scattered, taste and smell alone, more fragile but more enduring, more unsubstantial, more persistent, more faithful, remain poised a long time, like souls, remembering, waiting, hoping, amid the ruins of all the rest; and bear unflinchingly, in the tiny and almost impalpable drop of their essence, the vast structure of recollection.

Shepherd and his daughter Kirsten Shepherd-Barr, an English professor at Oxford University, combined their talents to explore what was happening in Proust's narrator's brain. The madeleine, they wrote, is an ideal vehicle for flavor; the vapors of tea carry its volatile aromatic compounds through the retronasal pathway to the olfactory epithelium, where the receptors lie. A single one of those vanilla or lemon flavorings, with its unique molecular shape, might recall a fragment of the narrator's early memories; the brain then can use the fragment to summon the whole. This neural architecture helps to make flavor flexible and adaptable. Food is written into memories and emotions. The reverse is also true. As memories accumulate, they come back to shade flavor perceptions in the present. This is one way that flavor continually evolves.

• • •

Unlike smell, the sense of taste is less emotional than existential. The primitive wants and aversions that tastes generate are basic survival responses. The signals from taste receptors ping through the oldest parts of the cerebral anatomy, where instincts and urges play out. When they reach the neocortex, they are processed by the insula, which contains distinct regions where neurons fire to salty, sweet, sour, bitter, and umami. The insula appears obscure. In each hemisphere, it is hidden away beneath a layer of tissue called the operculum, tucked into the brain's cortical shell at the temple. But in fMRI studies, it pops up again and again as a critical node in networks of brain activity for many different things. It seems to shape the overall tone of experience itself.

The insula seems to be where the body's internal state and external circumstances are sorted, assessed, and relayed to consciousness. Along with tastes, it processes other messages about the state of the body such as thirst, sexual arousal, temperature, the metabolic and cardiovascular stresses of exercise, and the need to use the bathroom. It aids in tasks involving perception, including the ability to distinguish one's own face in the mirror from that of someone else or a scrambled image; keeping rhythm to music; and processing emotions such as sadness, happiness, trust, empathy, beauty, and "state of union with God." It activates when we're engaged in sophisticated tasks, such as keeping time, recognizing an image being revealed piece by piece, or making choices. The insula, in other words, helps create the special, ever-shifting quality of now.

The unity of taste and smell in flavor is like a good marriage. The differences are profound, but each partner has complementary strengths and weaknesses. Their paths through the brain unite—along with those of all the other senses—in the orbitofrontal cortex, located above the orbits of the eyes. Relative to body size, humans have larger orbitofrontal cortices than any other animal—this was one of the most important evolutionary upgrades in the emergence of *Homo sapiens*. Flavor is only one in its array of sophisticated cognitive responsibilities, which includes decision making. It's the brain's food critic, connecting to areas governing emotions and judgment, and anatomically structured to process pleasure and aversion. Moving from the center out, its pleasure-sensitive neurons give way to displeasure-sensitive ones. This may explain our tendency to rank favorite or most hated foods: our brains are literally organized that way.

But the core of flavor perception is the way that the orbitofrontal cortex weaves the senses together, and with them all the elements of flavor. Thus, we perceive dark chocolate, grilled fish, and absinthe, rather than the long lists of distinct tastes and smells that comprise them. Individual tastes and aromas work in concert, reinforcing each other, fusing into something new.

The umami taste plays a special role in this process. Cooking, curing, and fermenting release prodigious quantities of umami, which dominates in seared meats, cheeses, tomatoes, pickles, and especially Asian foods such as soy and fish sauces and miso. Umami receptors detect glutamates—the salt of a particular amino acid, amino acids being one building block of proteins. (Umami is often dubbed the "protein taste," as sweetness is the taste for sugars, though its exact purpose is unclear. In nature, proteins are found mostly in milk and in raw animal flesh, which aren't very savory.) The surge in umami from the new foods of civilization was a nutritional as well as a taste bonanza. Glutamates fuel digestion and make it possible for the neurons in our brains to fire. In pregnant women, the placenta uses glutamates as an energy source. Umami receptors are found not just on the tongue but also lining the small intestine: extra glutamates stimulate better digestion and absorption of nutrients.

The Japanese word *umami* is a combination of the characters for "delicious" and "taste." Its meaning conveys a sense of wonder and satisfaction from food. Yet the umami taste is elusive. A sip of pure, dissolved glutamates is practically tasteless. But in concert with other flavors, umami's savoriness comes alive, and brain scans show it sparks patterns of brain activity roughly similar to those of sugar. The other four tastes boldly announce themselves, but umami operates by misdirection, helping other flavors bloom. It's like the Wizard of Oz, putting on a tremendous show from behind a curtain.

Two neuroscientists at Oxford, Edmund Rolls and Ciara McCabe, explored this phenomenon in a 2007 experiment. They treated twelve volunteers to an umami cocktail (monosodium glutamate—the chemical version of umami that food companies and Asian restaurants use—and a second substance, inosine 5'-monophosphate, which enhances MSG's effects) and a vegetable aroma. Separately, the drink and aroma were unpleasant. Together, their flavor was delicious. To map this curious effect, Rolls and McCabe tested the volunteers using fMRI scans. The umami-aroma combination ignited neurons in the orbitofrontal cortex in far greater numbers and for a longer time than one would expect from simply adding up the separate effects. Umami taste worked in concert with smell to create a new and more powerful sensation.

This suggests an underlying reason why the taste of chicken soup or pizza is delicious: umami unites and heightens taste and smell, adding a burst of pleasure. Think of what Parmesan cheese does to enrich the flavors of pasta it's sprinkled on: it produces a Technicolor burst of boldness and robustness. For ancient peoples sampling cheeses or fermented soy, this effect must have been a revelation.

Beverages also offer potent fusions of taste and smell. Ethanol, the alcohol in all alcoholic beverages, is a promiscuous molecule. It affects the brain's taste, smell, and touch systems all at once. These merge into its powerful effects on mood. Take a sip of wine, beer, or bourbon, and the alcohol binds to sweet and bitter receptors, and to the heat-sensing receptors that trigger the burn from chili peppers. Depending on the strength of the drink, any one of those sensations can move to the foreground. Below a 10 percent concentration, alcohol creates a faint sweet sensation, and the brain's response echoes that for sugar. This isn't surprising, as yeasts feasting on sugars churn out ethanol molecules. It's also influenced by genes: people with a family predilection for sweets tend to drink more.

But in stronger drinks, bitterness and burning overwhelm the sweetness, which is why distilled spirits like vodka or tequila, concentrated at 40 percent alcohol or higher, have such a kick, the edgy mixture of pleasure and aversion that makes tossing back a shot so bracing. As ethanol molecules evaporate, they waft into the nose and attach themselves to the receptors for smell, accounting for the jolt in a mere sniff of absinthe.

Alone, ethanol is relatively flavorless. It serves as a scaffold for other by-products of fermentation. Some of these substances pique the taste buds, some operate via smell. Jiahu grog contained acids and bitter compounds that balanced out its sweetness. Tannins, chemicals found in grape skins, create a distinct puckering sensation and bind to proteins on the tongue to alter the composition of saliva. Science still has only a sketchy idea of how most of these flavor compounds work; tracing the connection between fleeting molecules and flavor is a daunting task. One class of aroma molecule found in cabernet grapes, methoxypyrazines, imparts a fresh vegetable-like flavor similar to bell peppers.

Taste and smell blend so seamlessly in flavors that the different senses merge, becoming indistinguishable. The brain even mixes them up: in the mind, smells become tastes. Vanilla, an aromatic flavoring, is usually perceived as sweet. In a study, a majority of volunteers described the scent of strawberries and amyl acetate, a banana-flavored food additive, as sweet. Food formulators routinely add such aromatic essences to drinks to enhance their sweetness without using sugar. But this is a trick of perception: smells cannot be sweet. Sweetness is a taste, detectable only by receptors on the tongue. Somehow, the brain produces the sensation that the nose is tasting, or that the tongue is smelling—or both.

What's going on here? This sensory confusion resembles synesthesia, a neurological condition once considered a sign of genius or madness. Flavor may even be a form of it. In the brains of synesthetes, one sense triggers another, seemingly unrelated one. One to two percent of people have the most common form, seeing colors associated with words or symbols. Synesthesia involving taste or smell is relatively rare. In one century-old case, a man felt colors appear on his tongue and the inside of his mouth when he ate. In a more recent one, when a man ate, the flavors produced the sensation of three-dimensional geometric shapes; he could both see them and feel their contours with his hands. As the taste experience changed over time, so did the shapes.

In 2003, a pair of scientists, Jamie Ward at University College London and Julia Simner of the University of Edinburgh, ran a series of tests on a middle-aged British businessman, known by the initials JIW, with a rare form of synesthesia. Starting around the age of six, certain words and sounds triggered flavors in his mouth. This happened when he heard, spoke, read, or simply thought them, making it a constant source of irritation: it was hard to read a book, or concentrate during meetings. Flavors sometimes lingered on his tongue indefinitely until a new word-taste came along to displace them. His dreams were filled with tastes.

The way JIW's brain mixed up sounds and flavors roughly parallels the more common confusion of taste and smell. The scientists wanted to trace his sensations back to their point of origin in the brain, in hopes of isolating the connection between the two senses, and the nature of these mix-ups themselves.

There are two main theories of how synesthesia works. The first says that it's a remnant of infancy. Babies are born with an astonishing number of cross-connections in their brains, so our earliest sensory experiences are a fugue of feelings, sights, sounds, touches, tastes, and smells. Over time, many of these redundant connections are "pruned" by learning and experience. The result is a mature set of separate senses. Sometimes, though, a screwy connection remains intact, a lingering echo of an infant's experience. It's a hardwired channel that's not supposed to be there. It reaches across the brain, linking two distinct systems together, sending unwanted messages back and forth.

Of course, the senses already converge in one spot—the orbitofrontal cortex, where higher cognitive functions assemble and evaluate them, usually without confusion. The second theory suggests the sensory muddle occurs here.

To test these theories, Ward and Simner made a list of JIW's food associations, which were worthy of a novelist. The word "this" tasted of "bread soaked in tomato soup." "Safety" was "toast, lightly buttered," and the name "Phillip" evoked the flavor of "oranges not quite ripe." The associations seemed random at first, but when the scientists broke individual words down into smaller units of sound, some triggered tastes apparently linked to the meaning of the word: "blue" produced an "inky" flavor, "Virginia" the taste of vinegar. "Human" evoked the taste of baked beans, maybe the result of the missing "being." Forty-four words, including "cabbage," "onion," and "rhubarb," summoned the flavor of the food they described.

These results suggest that theory number two is correct, at least in JIW's case. Words are a form of abstract knowledge, and their meanings, not just their sounds, triggered flavors in JIW's brain. This was evidence of crossed connections among cognitive functions, language, taste, and smell. This link could not have formed during infancy, only after JIW learned to speak (though it was not clear why it happened). On some level, JIW's synesthesia was learned, shaped by his experiences. This shines some light on how the brain merges and confounds different tastes and smells: it learns to do so. As diverse sensations pop up repeatedly in pleasurable combinations, links form between them, down to individual neurons in the orbitofrontal cortex. In the brain, sweet tastes and "sweet" smells converge, making the sensations indistinguishable.

Fermentation's powerful, evocative sensations transformed it from a mere culinary technique into a cultural force. Hints of this transition appear in myths dating back thousands of years. Homer incorporated them into his epics, the first written Western literary masterpieces. They were composed around the end of the eighth century BC, probably in an area that is now part of modern Turkey, not far from the birthplace of herding and cheese. In one of these stories, Odysseus and his crew, trying to make their way back to Greece, arrive at the island of the Cyclopes, giant, one-eyed monsters. They sneak into the Cyclops Polyphemus's cave while he's off with his flocks and help themselves to the large, flat cheeses they find carefully arrayed on racks. When Polyphemus returns, he curdles milk to make more cheese. But soon he discovers his stores have been raided, finds the Greeks, and starts eating them one by one, devouring "entrails, flesh, and bones, marrow and all." Odysseus devises a plan to get the Cyclops drunk on a bottle of fine wine and blind him with a sharpened stake. He and his remaining crew escape strapped to the undersides of the rams when the wounded Cyclops, still a responsible herdsman, lets them out to graze.

The Cyclopes are on the cusp between savagery and civilization. They live in caves and eat men raw. Their island would make ideal farmland, Odysseus observes, but it's wild and overgrown. But they do have some glimmers of sophistication. They make wine from wild grapes, herd goats and sheep, and make cheese. This cannot fully redeem the Cyclopes in the eyes of the Greeks, but that doesn't stop them from tasting the cheese.

CHAPTER 5

The Seduction

Chef Homaro Cantu sometimes stages "flavor tripping" demonstrations for guests at his Chicago restaurants. Four slices of lime, six slices of lemon, and two plastic Japanese-style soup spoons—one holding a dollop of sour cream, the other plain Greek yogurt—are arrayed on a table. A plastic tub containing a fluffy, salmon-colored paste sits nearby. Participants are instructed to put a scoop of the paste in their mouths, allowing it to settle on the tongue. It's cool, with a slightly sweet flavor, pleasant but bland. After a few minutes it melts away. Then the tasting begins.

The paste is made from an extract of the miracle berry, from the West African shrub *Synsepalum dulcificum*. The red berries contain a protein called miraculin that plays an unusual trick on the sense of taste. Alone, its molecules block sweet receptors from doing their job: the taste of sugar loses its sweetness. But in the presence of acids, miraculin ignites them. The more acidic a food is, the sweeter it will taste. Since many foods contain acids, including fruits, vegetables, cheeses, and spices such as black pepper, this effect temporarily transforms flavor. The taste of lemon becomes light, delicate, like lemonade but less cloyingly sweet. Limes taste like oranges, yogurt like cream, and sour cream like cheesecake.

Cantu first encountered miracle berries while researching ways to make food more palatable for a friend whose sense of taste had been damaged by cancer treatments. Chemotherapy drugs travel in the bloodstream and saturate the saliva, causing a persistent bitter, metallic taste. Because taste bud cells are fast-growing like cancer cells, the drugs also target and kill them in great numbers. Radiation damages them, too. After testing various remedies, Cantu formulated his miraculin paste, which counteracted the metal taste. His friend was able to enjoy food again.

Cantu had established a reputation as an innovative chef at his first restaurant, moto, which opened in 2005 in Chicago's meatpacking district. He had played with diners' expectations, doing things like creating edible paper infused with flavorings; one of these was a picture of a cow that tasted like a freshly grilled steak. But his ambitions stretched beyond avant garde cuisine. He wanted to apply his culinary talents to social problems. The miracle berry, he believed, had hidden potential, which he set out to explore.

He worked with an importer to make miraculin pills, which dissolve on the tongue, available to thousands of cancer patients. He experimented with flavor effects, creating desserts designed to be tasted before, then after, a spoonful of miracle paste. He spent a week eating nothing but weeds, leaves, and grass from his yard, using the paste to make them palatable. Gradually, Cantu's attention turned to one of the worst food problems humanity faces. The taste for sweetness, an ancient and powerful urge once crucial to survival, has backfired in a spectacular fashion. The world is on a dangerous sugar binge.

• • •

The sweet taste is the body's signal that something biologically vital is at hand, proclaiming, "Devour me." Sugars are the foundation of the earth's food chain. Made by plants during photosynthesis, sugar molecules contain the sun's energy and make it available as fuel for all living organisms through their easily breakable chemical bonds. Because sugars are so useful, concentrated sources are rare in nature, found mainly in fruits, berries, figs, and honey. The combination of easy energy and scarcity makes sugars a prime target for hungry creatures, and sweetness a delicious and powerful motivator.

But humans found ways to overcome nature's limitations and make a lot of sugar, and feeding the world's voracious sweet tooth proved immensely profitable. Over the past three decades, two forms of sugar—crystals refined from sugarcane and sugar beets, and high-fructose corn syrup—have come to saturate diets as never before in history. Sugar flavors ubiquitous sodas, candy, and desserts. Corn syrup is added to many processed foods to enhance flavor, including bread, breakfast cereal, ketchup, baked beans, salad dressing, tomato paste, and even applesauce. Sugar has seemed to defy the law of supply and demand: the more that was added to foods, the more people wanted. Globally, the daily consumption of added sugars—those not already a natural ingredient in a food— rose 46 percent between 1983 and 2013, from 48 grams to 70 grams. Americans consume 165 grams per day, or 40 teaspoons, more than any other people in the world.

Humans evolved eating much smaller amounts of sugar; our bodies are not engineered to tolerate this much. A sugary diet can disrupt basic metabolic functions: how the body burns calories, stores fat, and processes nutrients. Over time, it can lead to chronic health problems including diabetes, obesity, cardiovascular diseases, and a lowered life expectancy. The flood of sugar in food closely tracks rising rates of diabetes and obesity: in 1980, 5.6 million Americans were diabetic, and approximately half the adult population was clinically obese. In 2011, 20 million had diabetes, far outpacing population growth, and three-quarters of adults were obese.

Sweetness, the most delightful of the basic tastes, has come to be seen as a menace to public health in the twenty-first century. A growing anti-sugar movement attacked food and soft-drink companies for using too much sugar and high-fructose corn syrup, and restaurants, movie theaters, and convenience stores for serving sugar-laden snacks. In New York City in 2011, then mayor Michael Bloomberg tried to limit the maximum size of soda cups to sixteen ounces, theorizing that smaller portions would cut consumption. But many people were outraged at what they saw as legislative overreaching, and the law was later overturned in court. Arkansas and West Virginia, among the states with the highest obesity rates, began taxing soda in hopes the extra expense would deter people from drinking it. Soda companies scrambled to find new sugar substitutes.

Cantu believed that none of these solutions would work. But he thought the miracle berry might. Unlike zero-calorie sweeteners such as aspartame or Stevia, it was not highly processed or refined. Its effects were pleasant and surprising. Cantu opened a restaurant named iNG in 2011 as a miracle berry showcase. Its eclectic tasting menus were built around flavor-tripping courses whose effects lingered through the meal. He changed course in 2014, closing iNG, and planned to make miracle-berry-flavored doughnuts and pastries the centerpiece of a new coffee shop, Berrista. These were steps along a path he hoped would lead to the mass market.

Cantu's ideas were fanciful, and he was competing against huge food and soft-drink corporations with substantial research and development budgets. But those obstacles were small next to the underlying problem: breaking the spell that sugar casts on the body and brain. The conscious perception of a taste seems like the whole point to us, but it's merely a decorative fillip on the cathedral of flavor. Underneath it lie the arches and buttresses that hold everything up: biological systems connecting flavor to the gut and the rest of the body. These connections infuse flavor with pleasure; they create cravings and compulsions and, for some, a dependency that resembles drug addiction.

How and why sugar came to seduce so many of us is a cautionary tale. Sugarcane, the world's primary source of refined sugar for thousands of years, is a species of grass. Its wild form must have frustrated prehistoric humans. Like a miser stuffing his mattress with cash, cane stalks store sugar in woody, indigestible cellulose fibers, using it to assist their own growth. Their husks can be stripped and the sugary insides chewed or sucked on like Popsicles, but it's difficult to get much sustenance this way. With the proper tools, they can be chopped, crushed, and boiled to produce only tiny amounts of crystallized sugar. Yet humans thought it worth the trouble. Along with bananas, breadfruit, and yams, they started to till sugarcane around 6000 BC in the region of modern-day Australia, Tasmania, and New Guinea.

A single artisan could brew beer or make cheese; making

sugar in large amounts required a higher degree of organization. A system of written knowledge, specialized workers, mills, boilers, trade routes, caravans, and ships sprang up around sugar in the ancient world. It was the ideal food product: delicious, and easy to transport without spoiling. The combination of economic cachet and deliciousness made sugar a catalyst for cultural and even spiritual change. Flavor became a force in history.

A story from about 2,500 years ago captures this foment. Two brothers were leading a merchant caravan of ox-drawn carts out of the town of Bodh Gaya in northeastern India when they noticed a man sitting beside the road. He was dressed in rags. Something about him caught the brothers' attention. "Stop!" they hollered back to the cart drivers. The brothers sent a boy to run back and dip into their stores.

The boy fished out a container of milk and some road food; accounts vary regarding exactly what it was. In some it's a knob of peeled sugarcane; in some, honey; in others a more stick-to-the-ribs concoction, rice cakes or sweet rice balls made with milk, honey, and molasses.

"Go ahead, eat!" the brothers yelled as the boy thrust the food at the man. They had a schedule to keep; an act of kindness could not take all day. But the man hesitated. Then he bit into the cake and smiled.

The man was Siddhartha Gautama, the Buddha. This incident took place a few weeks after his enlightenment. Buddhist scriptures say his insight, gleaned after a long struggle, freed the former prince from his desires: the cravings for food, sex, money, and success that cause the world endless trouble. Buddhism holds that all experience is tainted by cravings. For a while, Siddhartha had starved himself trying to extinguish them, but that had only made him crave food more.

Now, thanks to his enlightened state, Siddhartha apparently ate the sweet treat with no trace of these cravings, just simple enjoyment.

This account from ancient times captures a world grappling with this intense new sensation, whose pure taste and granular form made it preferable to honey. The Buddha lived in a sugarcane-growing region, and during his lifetime, India was starting to develop sugar refining into an industrial art, and created the world's first dessert cuisine. References to sugar started to appear in poetry, medicinal advice, and official records around the same time, including the *Arthasastra*, a governing manual written around 300 BC by a bureaucrat named Kautilya. He noted sugar's different forms in order of rising quality: *guta*, *sarkara*, and *khanda* (the second two are the roots of "sugar" and "candy"; *sarkara* is Sanskrit for "gravelly"). Members of the Jain sect, forbidden to kill even the tiniest living creature, could not eat honey because it might contain bee embryos. They turned to *matsyandika*, or sugar candy. Sugar was thought to keep the forces that ricocheted around the body in balance. Indian doctors believed eating it conferred special healing powers, helped digestion, and made semen more potent. According to an Indian book of cures from the second century BC: "In such a man's body even poison becomes innocuous; his limbs grow hard and compact like stone; he becomes invulnerable." One elixir of ginger, licorice, gum, ghee, honey, and sugar, if sipped each day for three years, was thought to guarantee a century of youth.

The two merchant brothers from the above tale, Tapassu and Bhallika, became the Buddha's first lay disciples: they continued to spread the Buddhist message on their travels. This reflects the later historical reality: to generate income,

Buddhist monks tended sugarcane and refined it. Over hundreds of years, both traders and Buddhist monks traveled the Silk Road, spreading sugarcane and the means for refining it.

But as sugar moved westward, it became an object of war. Early in the seventh century AD, the prophet Muhammad had founded Islam after receiving a divine revelation. He unified rival Arab tribes and territories into a growing empire spanning the Arabian Peninsula and beyond. Like the Romans before them, the Muslims of the Middle Ages didn't adopt other peoples' customs and technologies, they assimilated them. At this time in Persia, Sassanid-dynasty millers had figured out how to make pure granulated sugar. The name of the modern Iranian province of Khūzestān, still a major sugarcane-growing region, appears related to the word for cane, *kuz*, and *khuzis* for cane farmers. "Her lips aflow with sweet sugar / The sweet sugar that aflows in Khūzestān," read verses composed by twelfth-century poet Nezāmī Ganjavī. But Persia's geography, stretching northward, imposed limits on its mastery of cane cultivation. Sugarcane grows best in temperatures above 60 degrees Fahrenheit.

The Muslims had the climate and irrigation technology needed to cultivate cane, and the built-in trade routes that go with the spoils of conquest. In AD 642, only a decade after Muhammad's death, they conquered Persia, seizing its cane and its refining knowledge and technology.

"To enjoy sweets is a sign of faith," the Quran declares; its version of heaven is a garden with rivers of sweet water, milk, wine, and honey, mirroring the bodily humors. A third of the dishes in the *Kitab al-Tabikh*, a collection of more than three hundred recipes dating to ninth-century Baghdad (the time of Aladdin and his magic lamp, and other stories from

The Thousand and One Nights) are desserts. The Baghdad elite dined on forerunners of many modern treats, including ice cream, doughnuts, fritters, and pancakes flavored with sugar or doused in syrup.

The seeds of the modern, sugar-saturated world were planted during the East-West culture clashes of the Middle Ages. By the Third Crusade, late in the twelfth century (1,700 years after the Buddha), sugarcane grew from China to the southern Mediterranean and on to Morocco, yet sugar remained mostly unheard of in Western Europe. But French and English nobles and soldiers could not have missed it on their travels. In Sicily, where King Richard the Lionheart spent several months during 1190 and 1191, spiky stalks of sugarcane grew in broad patches on hillsides not far from foreign troop garrisons in Messina and Palermo, near sugar mills with steam spewing from their refineries. The plant had been brought there by Muslims two hundred years earlier. Sicilian sugar makers knew how to process large volumes of cane, and their product was used plentifully in the kitchens of the Sicilian nobility, and shipped throughout the Muslim world. When Richard's armies returned home—having failed to secure Jerusalem—they brought samples with them.

The earliest record of the word "sugar," taken from the contemporary Old French word *çucre*, is found in the accounting rolls of the Benedictine Abbey of Durham, in northeast England, from 1299, where monks kept track of various goods on hand, including "Zuker Roch" (rock sugar) and "Zuker Marrokes" (Morrocan sugar). Sugar wasn't considered a food, but a medicine, spice, and preservative. Twelfth-century theologian Thomas Aquinas wrote that eating it would not break a religious fast because it was a medicine taken to aid digestion. Over one year in the late thirteenth century, the royal household of King Edward I used nearly a ton of sugar flavored with rose petals, a common cure for various illnesses—far more than the 677 pounds it used for food. As late as the eighteenth century, Carl Linnaeus, the Swedish scientist and classifier extraordinaire, named the most common sugarcane species *Saccharum officinarum*, meaning "sugar from the apothecary." Medieval European doctors embraced sugar cures touted by both the Arabs and the Byzantine Greeks. A popular Arab treatment for the common cold was *al fanad* or *al panad*, small sugar twists made from congealed syrup, which became known in English as alphenics or penides. In 1390, the Earl of Derby paid "two shillings for two pounds of penydes," the Oxford English Dictionary recounts: the first cough drops.

The modest size of this culinary niche was a function of geography. Most of Western Europe was too cold to grow cane, and trade only brought so much. Europeans solved this problem in the traditional way, by conquest. But over time, they added new elements: capitalism and the Industrial Revolution.

On his second voyage to the New World in 1493, Christopher Columbus brought sugarcane from the Canary Islands—the farthest west it was then found—to plant on Hispaniola. It was a prescient decision. Tales of vast caches of gold and silver on the island turned out to be wild exaggerations. Making sugar became the only reliable source of wealth. Hispaniola's warm climate, similar to that of sugarcane's primordial birthplace near New Guinea, was ideal, and space unlimited. Sugar even interested the Taino natives, who began planting their own.

Spain's colonies were expected to produce or die. Gonzalo de Velosa, a colonist trained as a surgeon, was canny enough to see that sugar production wasn't for amateurs. It was now traded in the global marketplace, and in Europe the price was rising; a real investment might pay off. So in 1515, he paid a small fortune to bring experts from Canary Islands sugar plantations to Hispaniola. They built mills that could be powered by horses, cattle, or waterwheels, which could produce far more sugar than mills worked by men. By 1520, six sugar mills were operating and another forty being built. But then these budding sugar entrepreneurs found their labor force was dying. Many Taino fell ill from infectious diseases they contracted from the Europeans; more died from forced labor. To fill this void, the Spaniards began to import slaves from Africa.

In 1660, a young man named Thomas Tryon set sail from London for the British colony of Barbados, then a growing center of the New World's booming sugar industry, joining thousands of Englishmen then risking everything to move across the sea to get rich. Tryon was not an entrepreneur, however, but a seventeenth-century flower child. He hated the excesses of the modern world and fancied himself a philosopher of food, out to foment a peaceful revolution that would tame cravings and gluttony and make his followers right with God. Apprenticed to a London hat maker as a young man, he dabbled in herbalism, magic, and alchemy. He also tried silence and asceticism, starving himself to observe his body's reactions.

The New World captured Tryon's imagination. To most, it symbolized paradise, a Garden of Eden, wild and pure. Tryon saw it differently: a place where the natives lived in perfect harmony with nature. The reality shocked him. The hillsides of Barbados had been stripped of trees, bushes, and brush; stands of sugarcane rose in their place. Twenty-five thousand

African slaves worked this giant sweetness machine. During the harvest, they roamed the fields with machetes, slashing down stalks, stacking and hauling piles of cane weighing hundreds of pounds. In makeshift sheds, water-powered mills with three giant rollers set on end crushed cane stalks fed into them by sweating laborers. The work was tricky: if a man slipped up, fingers or limbs and sometimes entire bodies might be pulled into the rollers and crushed; an ax was kept handy for emergency amputations. Workers lugged buckets of juice to the boiling house, where dried-out cane stalks and other detritus were kept burning around the clock under copper kettles. A master refiner kept an eye on the pots as they boiled and filtered the soupy brown syrup into smaller kettles. To keep the sheds from bursting into flame, the roofs were periodically doused with water. Finally, crystallized sugar was dried, packed in 1,500-pound barrels, and hauled by donkeys to Bridgetown harbor. There, sugar barons, merchants, and servants mixed with vagrants, ex-prisoners, and rum-swilling lowlifes. Yellow fever epidemics routinely claimed hundreds of lives; bodies were thrown in a bog at the edge of town, which belched horrible odors that hung in the air.

Tryon spent five years in Barbados as a milliner before returning to London. His experiences continued to preoccupy him. Nearly twenty years later, he began to write pamphlets espousing a passionate, proto-vegan philosophy. He denounced the sugar plantations as monuments to greed and gluttony, and sweets as temptations that obstructed the digestive tract. Like the ancient Greeks, he believed that the basic tastes of salty, sweet, sour, and bitter defined a man's character. Tastes "can readily open all the doors and secret gates of

Nature's Cabinets," he wrote later; the flavor sense was "the prince, king, or complete judge over Life and Death." Overindulging the appetites ultimately led to damnation. Tryon's words and arguments influenced the abolitionists (though he was too much a traditionalist to argue for the abolition of slavery). His views shaped an entire vegetarian movement: Ben Franklin gave up meat in his youth after reading Tryon's writings.

But Tryon was fighting inexorable forces. The foundation of a global sugar behemoth had been laid. From the seventeenth century on, rivers of granulated sugar flowed from Europe's Caribbean and South American colonies to the larders of kings and, for the first time, into the homes of the middle and lower classes. Food grew sweeter. French chefs began inventing sweet dishes—pies, mousses, pastries, and puddings—that were now separated from the main courses and served as a meal's finale. The word "dessert," first used in the seventeenth century, comes from the French word *desservir*, literally to "un-serve," or to clear the table when dining concludes. In Tryon's England, desserts were still regarded as a French abomination, but a few decades later they became standard fare. People began sweetening imported drinks previously taken straight: hot chocolate from the New World, coffee from Africa, and tea from China. In Britain, annual per capita sugar consumption grew from four pounds in 1700 to eighteen pounds in 1800, and then to ninety pounds in 1900.

Between bakeries and rum distilleries, Big Sugar became a major employer, and was soon part of the bedrock of English culture. Near the end of his life, Thomas Tryon was finally seduced. He endorsed sugar, and the plantations he had once condemned. "It spreads its generous and sweet influences through the whole nation," he wrote. "And there are but few eatables and drinkables that it is not a friend to, nor capable to confederate with; and upon the whole, there is no one commodity whatever, that doth so encourage navigation, advance the King's customs, and our land, and is at the same time of so great and universal use, virtue and advantage as this King of Sweets."

The Buddha's humble snack had embodied moderation and balance in a world where temptation was growing. With Tryon's conversion, excess had won. In the three centuries that followed, the sugar system expanded inexorably. After the British blockaded France in 1806, halting sugar imports, Napoleon turned to an obscure white root grown and refined in a small part of central Europe. Like cane, the sugar beet contained sucrose, but unlike cane, it grew in cold climates. Napoleon invested a million francs to train farmers and refiners, and another global sweetness industry was born. More than a century later, in 1957, two scientists working for the Corn Products Refining Company of Argo, Illinois, found a way to convert the glucose that makes up cornstarch into far-sweeter fructose, creating high-fructose corn syrup. The United States is the world's biggest corn producer, and corn syrup can be pumped directly into food factory vats. By the 1970s, corn syrup became a standard food additive.

The world is now running a giant sweetness experiment. Enormous resources are employed, year after year, to saturate billions of living bodies with sugars, and scientists are only beginning to assess the effects on human biology and public health. What is it about sweetness that exerts such an irresistible attraction? To put the question more broadly: What makes any food tasty, and why? What biological purpose do such pleasures serve, and how do they slide so easily into overindulgence?

Sweetness is the most basic form of tastiness, and of pleasure itself. It is an ancient phenomenon. In evolutionary terms, it seems to be a force existing prior even to sex. Eons ago, single-celled organisms may have clumped together in order to consume more sugar faster, possibly the first event in the evolution of complex life. Fruit flies, whose ancestors branched away from those of humans during the Cambrian explosion more than 500 million years ago, have a similar taste for sugar and the behavioral programming to unerringly guide them to it. These primitive urges still echo in the pleasure *Homo sapiens* take from sweetness and anything tasty. Without it, meals would be lifeless collages of flat sensations.

Like the ancient Greeks, modern scientists often dismissed the pleasures of food, and pleasure generally. Early in the twentieth century, most believed it was feelings of discomfort that really mattered. They were what drove people to action: hunger led to eating, thirst to drinking, and lust to sex. Touch boiling water and the hand jerks back. The psychologist and philosopher William James summed up this thinking in a letter to a friend in 1901: "Happiness, I have lately discovered, is no positive feeling, but a negative condition of freedom from a number of restrictive sensations of which our organism usually seems the seat. When they are wiped out, the clearness and cleanness of the contrast is happiness. This is why anaesthetics make us so happy. But don't you take to drink on that account." In other words, the absence of discomfort is the only true pleasure.

In the 1920s, Sigmund Freud formulated a similar idea that primal drives push humans to seek sexual release. Twenty years later, behavioral psychologist Clark Hull devised the drive-reduction theory: when stressed or frustrated, a human or animal will take action to make the bad feeling stop, then try to avoid it in the future.

These assumptions all shared a rather dreary view of the human condition. They soon faced a challenge. In 1950, James Olds, a thirty-one-year-old psychology postdoctoral fellow at McGill University in Montreal, decided they were out of step with everyday experience. If chronic discomfort or pain was the key to all behavior, it implied that the best things in life meant nothing. He believed pleasure and happiness deserved to stand on their own.

"For an organism that seeks novelty, ideas, excitement, and good-tasting foods, the drive-reduction theory was a Procrustean bed," Olds wrote, referring to the myth of Procrustes, a son of the Greek god Poseidon, a rogue smithy who forced his guests to fit in his iron bed by amputating their limbs. "Whatever did not fit was shorn from our image of the man and the rat. Drugs, good foods, and sex were thought of in terms of a need—that is, a hurt generated by withdrawal."

At the time, it wasn't clear to anyone how the brain produced a sensation of tastiness or gratification, or why. Olds set out to invent a science of pleasure. He'd had his initial insight while doing an experiment with an albino lab rat. The rat was in a specially designed box, three feet square, its sides a foot high, called an operant conditioning chamber, or a "Skinner box," after its inventor, B. F. Skinner, a founder of the school of behaviorism early in the twentieth century. Behaviorism was an alternative to Freudian psychology, with its focus on hidden motivations. Skinner believed it was more scientifically rigorous to leave the mind out of it. His box reduced behavior to its essence: an animal was placed inside it, a stimulus introduced, and the response observed. The typical stimulus was either a punishment, such as a mild electric shock, or a reward, such as sugar water. But Olds had found a way to skip over those and go straight to the place in the brain where pleasure and pain were thought to form.

Working with his colleague Peter Milner, Olds had surgically threaded an electrode into an area in the rat's brain near the hypothalamus. The wire ran to the ceiling, where it connected to a stimulator activated with a button. Pressing it would instantly trigger some kind of reaction in the brain—whether a flood of pleasure, a jolt of pain, or some other sensation or emotion, Olds didn't know.

He decided to stimulate the rat each time it entered a corner and see how it responded. The first time, the rat circled back to the same corner; it seemed to like the stimulation. Olds hit the button again. The rat returned to the corner much faster. The third time, it stayed put, waiting expectantly for more.

At first, Olds thought he might have discovered the source of curiosity; the rat returned to the corner because it was intrigued. But when he modified the box so that the rat could stimulate itself by pressing a lever, it displayed no adventurousness at all; it just sat and pressed repeatedly. The electrode seemed to make the rat feel good. This effect proved very, very powerful. The rats in Olds's experiments ignored sugar solution, food, water, and the chance to mate in order to press the lever. In one experiment, they pressed until they nearly died of starvation and thirst. In another, they ran across the floor of a box wired to deliver shocks to their feet in order to hit the switch. He positioned electrodes in slightly different places to map presumed pleasure-related areas in the brain. Stimulating one area made rats eat ravenously, while another made them lose all interest in food.

Eating and drinking proved the easiest behaviors to manipulate. "The 'rewarding' parts of the brain," Olds wrote, "were all related to olfactory mechanisms and to chemical sensors." Flavor and pleasure were, on some level, one.

Olds's discovery was dubbed the "pleasure center." It was a stunning advance, and scientists wondered if the same brain structures that drew pleasure from a spoonful of sugar might also be the source of sexual gratification, or the satisfactions of a lively conversation or finishing a good book. The media debated the potential advantages of this insight. Perhaps the terrible personal and social scourges of unhappiness and depression, not to mention the suffering that defined the human condition, could be cured with the flick of a switch.

But it wasn't quite that simple. In 1987, Kent Berridge, then a thirty-year-old junior faculty member at the University of Michigan, was working on an experiment with rats when he noticed something that bothered him. When rodents taste something sweet, their faces and mouths react in a characteristic manner, gaping a bit and flicking their tongues from side to side, as if licking their lips. This is their version of a smile, a clear outward sign of the inner experience of yumminess. The rats had been given a drug to shut down their pleasure centers, and became logy and indifferent as expected. But they also still licked their lips at the taste of sugar— apparently, they were enjoying themselves, though that was supposed to be impossible. At first, Berridge shrugged this off as probably trivial.

Berridge's grinning rats had been given a drug to block a powerful brain hormone called dopamine. In the years since Olds's experiments, it had been identified as the chemical that powered the pleasure center. Dopamine is a neurotransmitter, a hormone that the brain employs to send messages in concert with firing neurons. Neurotransmitters facilitate everything from movement to emotions. In Olds's day, dopamine was an obscure brain chemical, thought to be a building block of more important hormones, adrenaline and noradrenaline, with no apparent function of its own. Scientists first grasped its importance in the 1960s when they discovered that it was essential to voluntary movement—in fact, it's the dying off of dopamine-generating neurons that leads to the tremors and paralysis of Parkinson's disease. Biologist Roy Wise later found that rats given dopamine-blocking drugs displayed precisely the opposite effects of a pleasure electrode. The rats slumped into utter indifference. They stopped eating and drinking; sweetness and all other pleasures lost their allure.

Wise proclaimed dopamine the pleasure chemical, and the scientific community followed suit. "The dopamine junctions," he wrote in 1980, "represent a synaptic way station . . . where sensory inputs are translated into the hedonic messages we experience as pleasure, euphoria, or 'yumminess.'"

Berridge reran his rat experiment. The results were the same. So he began to search for an explanation for why dopamine-free animals could still savor the taste of sugar. He wondered if Wise was wrong. (The two of them were collaborating at the time, so it was a bit awkward.)

One obstacle he faced was that, beyond its facial expressions and behavior, a rat cannot explain how it feels. While researching old pleasure electrode experiments, he found an intriguing way around that problem. Between the 1950s and 1970s, doctors at Tulane University in New Orleans had implanted electrodes in the brains of human volunteers. Most had severe forms of mental illness; researchers hoped that brain stimulation would alleviate their symptoms. (Today, a more exacting variant of this technique, deep brain stimulation, is used to treat severe depression.)

The experiments were revealing, helping psychologists map the sources of behavior and emotions in the brain's anatomy. But they were sometimes spectacularly wrong-headed. In one, psychiatrist Robert Heath implanted nine electrodes in the brain of a young man, code-named B-19, who was severely depressed and had not responded to either drugs or talk therapy. He was also gay, and one aim of the treatment was to "cure" him; therapies included viewing a stag film, and a two-hour visit from a female prostitute.

With so many curling wires dangling from his skull, B-19 looked like a cyborg, and in a way he was: he became a kind of electronic puppet pulling its own strings. Heath gave him a button to activate the electrodes. One was placed in the pleasure center. And sure enough, when a small jolt was administered there, B-19 acted just as the rats had. He kept punching the button: during one three-hour period he hit it 850 times. He reported a strange mix of feelings: self-confidence, relaxation, and arousal. When the lab technicians started to disconnect him, he begged them not to. The electrode also made him want to have sex with both men and women, leading

Heath to think he'd found a potential cure for homosexuality. After several weeks of experiments, the electrodes were removed and B-19 was released. Heath followed his progress for eleven months. "While he looks and is apparently functioning better, he still has a complaining disposition which does not permit him to readily admit his progress," he wrote. Following his release, B-19 held a number of part-time jobs, had a ten-month sexual relationship with a married woman, and told Heath he had twice turned tricks with men to make money.

Reading these descriptions, Berridge noticed something. The electrodes were assumed to be stimulating eruptions of dopamine in B-19's brain—yet he never seemed to enjoy himself. He became sexually aroused, but never had an orgasm. He never said "oh, that feels good!" Hitting the button led only to more anticipation. Perhaps dopamine did not really create pleasure after all, Berridge thought, but rather the craving for it. Once, scientists had dismissed pleasure's importance. Now, they might be mixed up about what caused it.

Searching for alternative pleasure chemicals, Berridge looked to addictive drugs. Opioids such as morphine and heroin evoke feelings of euphoria. Perhaps the answer was the brain's own natural opioids, also known as endorphins. In the early 2000s, nearly two decades after his initial discovery, he tracked intense pleasure reactions to endorphins in two areas of a rat's brain, the nucleus accumbens and ventral pallidum. He named these "hedonic hotspots." These tiny clumps of neurons, about the size of the head of a pin, are the only brain structures known that directly cause pleasure.

The neurons in hedonic hotspots respond to several different endorphins, suggesting that pleasure is complicated, the result of many brain systems interacting at once. One such endorphin is orexin, a comparatively rare substance also connected to appetite, arousal, and wakefulness; another is anandamide, named for the Sanskrit word *ananda*, which means "bliss." It plays a role not only in pleasure but pain, memory, and higher thought processes. Orexin and anandamide activate opioid and cannabinoid receptors, respectively, which also respond to heroin and marijuana.

The anatomy of pleasure bridges the visceral and the brain's higher functions, placing hedonic hotspots smack in the middle. They work something like circuit boards. There are two hotspots, as well as a "coldspot" nearby that sparks disgust. The coldspot is nestled in an area rich in dopamine neurons that inspire intense wanting: when the two were stimulated together, Berridge made a rat yearn for something that tasted terrible. Removing one hotspot reduced pleasure but didn't eliminate it, but removing the other made sweet things taste terrible. That could mean this hotspot's job is to inhibit disgust and enhance pleasure at the same time.

The simple delight of sugar dissolving on one's tongue appeared to be the product of certain neurons tucked into structures deep inside the brain percolating in a cocktail of the body's most intoxicating hormones. Yet however detailed these anatomical maps of the origins of pleasure became, they could not explain its purpose. The role of the cravings caused by dopamine was also up in the air. As he learned more, Berridge formulated a theory he hoped would fill this void. Like many behavioral models, it was blunt, reducing the vast diversity of human decisions and actions down to the shape of a triangle.

The triangle's sides are named "wanting," "liking," and "learning." It can describe all behavior, but applies particularly to taste and to flavor. Wanting is a state of desire and heightened focus before food is eaten. Liking is the pleasure of a good taste, a reward for doing the work of obtaining food. Wanting and liking work in tandem to forge learning. The human brain very quickly picks up on how to gratify itself, learning where the tastiest food is and how to get it.

In the 1990s, Cambridge University neuroscientist Wolfram Schultz did a series of groundbreaking experiments that dramatized this dynamic. Schultz also showed that dopamine was the hidden hand in cravings: it is what powers "wanting." In one test, monkeys were placed in front of a computer screen that displayed geometric patterns. One pattern flashed two seconds before sugar syrup was dispensed from a bottle; the other appeared randomly. Electrodes measured the activity of a single dopamine neuron in the monkeys' brains. At first, this neuron fired when a monkey sipped. But as the cycle repeated and the monkeys picked up on the signals, the neuron adapted. It began to fire *before* the treat arrived— predicting a good taste was coming and sharpening the anticipation and craving for it. When kitchen smells make the mouth water, that's dopamine setting the sensory table. And if learning can be tracked in a single neuron, imagine billions of neurons in the human brain doing the same over the course of a lifetime.

Having identified building blocks of pleasure, Berridge pondered what the fleeting "goodness" in a sweet taste really was. Clearly, it was distinct from the feelings that accompany listening to a favorite song, or seeing an old friend. But deep down, these states might be the same—formed in the same areas of the brain, reliant on the same patterns of firing hedonic hotspots and hormonal flux. Evidence from fMRI scans suggests there's something to this idea—the different forms of pleasure have patterns of brain activity that closely overlap. As humans evolved and culture made its imprint on the human brain, perhaps the ancient neural circuitry responsible for sweetness was adapted as the template for more exalted pleasures, and maybe even happiness itself. "Final happiness may be a state of liking without wanting," Berridge said. "That may be a Buddhist sense of happiness."

Overindulging in sugar disrupts the normal rhythms of wanting, liking, and learning. Humans evolved eating just enough food to sustain big brains and lithe, active bodies. The stomach can hold only so much, and the gut and brain engage in a continual dialogue to ensure a balance is struck. Powerful hormones excite the dopamine-sensitive parts of the brain, spurring humans to seek food when hunger strikes. Pleasure peaks at the start of a meal, when hunger is keenest, then declines—one reason why no one eats the entire contents of a sugar bowl. But persistently overdose this system, and the signals start going awry. Fructose appears to raise levels of the hormone ghrelin, for instance, which stimulates hunger; instead of satisfying, eating sugar leaves us wanting more.

Science has only begun to trace these easily corruptible pathways running between body and brain. Thanks to our expanding knowledge of taste genes, lab mice and rats can be genetically engineered with specific genetic traits for experiments. Ivan de Araujo, a neuroscientist at Yale, fed both plain water and sugar water to a type of mouse engineered to have no taste for sweetness. They should not have been able to sense the difference, yet they strongly preferred the sugar water.

Ordinarily, the tongue's sweet receptors signal the brain that a delicious reward is on the way. With that signal absent, de Araujo suspected that the sugar was still making its presence known through an unknown back channel, making the mice crave sugar with no conscious awareness of it. To test this hypothesis, he implanted probes in the mice's brains that measured their dopamine levels; the sugar water produced a dopamine rush. Somehow, de Araujo thought, the body sensed the sugar—perhaps through taste receptors lining the gut—and signaled the brain it was there, triggering the yen for more. When the experiments were repeated with human subjects—their sweet taste blocked with a drug—they described a hazy sense of satisfaction after sipping sugar water.

These urges resist both willpower and medications. Appetite suppressants reduce hunger, but craving and pleasure are more complicated phenomena. Dopamine-blocking drugs shut down the hankering for sugar, but extinguish all motivation at the same time. A drug aimed at suppressing the pleasure of food might kill all joy along with it.

As more of sugar's insidious effects were discovered, people began getting the message. In the first few years of the 2010s, sales of soft drinks—the single largest source of dietary sugar in the United States—leveled off, and then decreased for the first time in their century-long history. Overall high-fructose corn syrup consumption fell, too. Obesity rates plateaued, though at high levels nutritionists still found alarming. Diabetes rates, however, continued to rise. It will take years to assess the public health toll.

The ideal solution would be a sugar substitute that perfectly mimics the taste and poses no health risks. But this is the oldest unsolved taste problem in the world. The Romans boiled crushed grapes in lead vessels to make a syrup called *sapa*, used to sweeten wine, stews, and other dishes. The active ingredient was lead acetate, also known as "sugar of lead," created by a chemical reaction between the grape juice and the containers. It was also toxic. Some have claimed Rome fell because its entire ruling class suffered from *sapa*-induced lead poisoning (historians are skeptical of this explanation). Lead acetate was still used for centuries afterward as a wine sweetener; among its possible victims were wine-drinkers Pope Clement II, who mysteriously dropped dead in 1047, and, eight hundred years later, Beethoven.

Modern sugar substitutes such as saccharin, the active ingredient in the pink packets of Sweet'N Low, and aspartame, used in diet soda, have their own problems. They don't taste like sugar. Table sugar, made from sugarcane or sugar beets, is made of sucrose, a molecule in turn made of two sugars, fructose and glucose. High-fructose corn syrup is a physical mixture of those two, with slightly more fructose. Of all the sugars, fructose is the sweetest. The uncannily precise bond between sweet receptor and fructose molecule means there is no other substance capable of exactly mimicking its taste.

The molecules of substitutes bond to sweet receptors, but don't fit perfectly, like a key that slides into a lock but won't turn all the way. They also bond to other types of receptors, including those for pungency and bitterness. The result is odd, off flavors, such as aspartame's faint metallic aftertaste, that don't fully ignite the brain's pleasure circuitry. Most sugar alternatives don't dissolve well in water, either, and will cling to the tongue rather than remain in solution. This means they pack a sensory punch—aspartame is about two hundred times as sweet as table sugar—but it can also make their tastes linger too long. The differences in chemical structure also make them poor baking ingredients. Sugar is not just sweet but versatile. Heat it and complex flavors emerge, with hints of acidity and bitterness. It can take multiple forms and consistencies, from crystals to caramels, that no substitute can touch.

Today's dominant artificial sweeteners are also lab-engineered industrial chemicals. Saccharin is a coal-tar derivative accidentally discovered at Johns Hopkins University in 1878. Aspartame was found in 1965 when a lab scientist at the pharmaceutical company Searle absentmindedly licked his index finger, which was dusted with an ingredient from an ulcer drug. Sucralose, the active ingredient in Splenda, was found while researchers for the sugar giant Tate & Lyle were studying ways to turn sucrose derivatives into an insecticide. The health concerns over sweeteners are more ambiguous than those surrounding sugar. Aspartame produces trace amounts of methanol in the intestine, a form of alcohol the body transforms into formaldehyde—the chemical used in embalming fluid, and a carcinogen—before it breaks down again. But so do oranges and tomatoes. The Food and Drug Administration banned saccharin in 1976 because of a tentative link to cancer in lab animals, but later unbanned it because the evidence was scant. Sucralose is not broken down in the body. However, recent studies ominously show that artificial sweeteners may contribute to diabetes.

Shell-shocked consumers are rejecting artificial ingredients on general principle. In 2013 alone, sales of Diet Coke and Diet Pepsi each plunged by 7 percent. Since the early 2000s, food and soft-drink makers have spent tens of millions of dollars in a race to find natural sugar substitutes. Many plants produce sweetish substances. But these don't taste exactly like sugar either. Thaumatin, a protein found in *Thaumatococcus daniellii*, a plant that grows in West African rain forests, is the sweetest substance known—three thousand times more potent than sugar. It lingers on the tongue for minutes, leaving licoricey aftertastes. Stevia, made from the leaves of the South American *Stevia rebaudiana* plant, has a bitter edge.

Homaro Cantu believed the frustrations of big food companies offered an opening for miracle berries, though he faced his own set of obstacles. In 1974, the FDA had classified miraculin as a food additive, meaning it would have to go through extensive testing before it could gain approval as an ingredient. Its advocates claimed that the US sugar industry, which wields substantial clout in Washington, had lobbied for this behind the scenes. Miraculin is currently classified as a dietary supplement. By the time Cantu found it, several startups had sprung up to market and sell miracle berry extracts. The price was still high—a single pill cost $1.50—but researchers had found ways to transfer miracle berry genes to tomatoes and lettuce, which can produce much larger amounts of miraculin than berries. Chemically, miraculin is not even a sweetener. Its flavor is mild. Instead, it alters other flavors, sometimes unpredictably. This may not be enough to start a diet revolution. But it does show that there are new frontiers of sweetness that have yet to be fully explored.

CHAPTER 6

Gusto and Disgust

As HMS *Beagle* sailed along the South American coast toward Tierra del Fuego in 1833, Charles Darwin had a series of scientific adventures. He had been recruited by the *Beagle*'s captain, Robert FitzRoy, as a geologist to aid the ship's principal mission, mapping the South American coastline and seafloor. In a few months, they would reach the Galápagos Islands, off the coast of Peru, where Darwin would find the strange flora and fauna on which he would base his theory of evolution. As the *Beagle* sailed south, he spent most of his time on land, observing and collecting geological specimens. At Bahía Blanca, Argentina, he rode with gauchos into the pampas and dined with them on roast armadillo. In Uruguay, he bought the skull of an extinct rodent, the size of a hippo's, from a farmer for eighteen pence. At Punta Alta, on the coast of Patagonia, he found bones from a megatherium, a huge, extinct armored sloth.

Darwin was both fascinated and repelled by the indigenous people living at the southernmost tip of the continent. He was twenty-three, on his first voyage, and had plunged into an alien world full of strange sensations. The people he met were the strangest of all. The Yahgan tribe lived a marginal existence as hunter-gatherers, roaming the archipelago near Cape Horn in dugout canoes. Most had long hair and wore little clothing, even in frigid weather. When the *Beagle* rounded the cape, he observed some of them rowing canoes. They struck Darwin as strange and degraded examples of humanity. "These poor wretches were stunted in their growth, their hideous faces bedaubed with white paint, their skins filthy and greasy, their hair entangled, their voices discordant, and their gestures violent. Viewing such men, one can hardly make one's self believe that they are fellow-creatures, and inhabitants of the same world."

Their food was vile. "If a seal is killed, or the floating carcass of a putrid whale is discovered, it is a feast; and such miserable food is assisted by a few tasteless berries and fungi," he wrote in his journal. The Yahgan would render the carcasses of beached whales, burying their meat and blubber in the sand. Without oxygen, it would ferment rather than decay. A few months later they would unearth it and feast. A shipmate who had spent time in the region filled Darwin in on the worst of it: cannibalism. When famine struck, the Yahgan ate their old women before their dogs. A native boy had explained the rationale: "Doggies catch otters, old women no." The unfortunate grandmas would sometimes escape into the mountains, then be captured and brought back to the hearth, where they were suffocated with smoke, then butchered for the choicest parts. (This was almost certainly a rumor; anthropologists have found no evidence that the Yahgan practiced cannibalism.)

On January 19, 1834, the ship anchored at the midpoint of the Beagle Channel (so named after the same ship explored it six years earlier), a one-hundred-mile stretch of water north of Cape Horn. A party of twenty-eight disembarked, including FitzRoy, Darwin, and three Yahgan, who had been captured on a previous voyage and were returning home after three years in England. They took four boats and rowed along the eastern bank. The following day, they entered an inhabited area; surprised Indians lit signal fires up and down the shore, and some followed the boats. The *Beagle* party came ashore near a Yahgan camp, and a tentative meeting took place. Initially hostile, the Indians warmed as the crew handed out gimlets (small tools for boring holes) and stretches of red ribbon that they tied around their heads. One of the three returning Indians, Jemmy Button, "was thoroughly ashamed of his countrymen, and declared his own tribe were quite different," Darwin wrote, "in which he was woefully mistaken."

As they all sat around campfires, Darwin opened a tin of preserved beef and began to eat. Canning had been invented only twenty years earlier, and canned meat had only recently become standard shipboard fare in the British Empire. Its taste was passable at best, something like the canned corned beef sold today, but a vast improvement over the smoked and salted meats in use a decade earlier, which rotted on journeys of any length.

"They liked our biscuit," Darwin wrote. "But one of the savages touched with his finger some of the meat preserved in tin cases which I was eating, and feeling it soft and cold, showed as much disgust at it, as I should have done at putrid blubber."

The next day, the party rowed to Wulaia Cove, where they left their "civilized" Yahgan companions, and continued to explore the area before returning to the ship a week later.

But the meat-tin incident stuck in Darwin's head. His observations of the Yahgan forced him to confront his prejudices. Unlike many educated Europeans of the time, he believed that the absence of civilization, not a savage nature, was responsible for the natives' bizarre tastes and abject circumstances. If this were true, the extremes he observed meant human behavior and sensibilities were even more malleable than he had imagined.

Almost forty years later, after *On the Origin of Species* had secured his place in history, Darwin wrote about this encounter in his new book, *The Expression of the Emotions in Man and Animals*. The book's main argument was a controversial, though logical, extension of his theory of natural selection: mankind's infinitely subtle emotional expressions, thought to be reflections of the soul, had evolved from those of animals. Both the Yahgan man's and his own reaction to the meat tin exemplified disgust, an emotion that had originated as a response to noxious foods but had evolved into something more complicated:

> The term "disgust," in its simplest sense, means something offensive to the taste. It is curious how readily this feeling is excited by anything unusual in the appearance, odour, or nature of our food. In Tierra del Fuego a native touched with his finger some cold preserved meat which I was eating at our bivouac, and plainly showed utter disgust at its softness; whilst I felt utter disgust at my food being touched by a naked savage, though his hands did not appear dirty.

Each man had been repelled not by the meat's taste or smell but by some ephemeral quality springing from a mixture of the sense of touch and imagination. For the Yahgan tribesman, it was the feel of this strange new food on his fingertip and the thought of it on his tongue. For Darwin, it was the idea of eating something touched by such a degraded example of humanity, who had, perhaps, consumed human flesh.

The word "disgust" comes from the Latin verb *gustare*, to taste and to enjoy; and the prefix *dis*, meaning "apart" or "not." It is literally deliciousness negated. Disgust is a uniquely human reaction based on ancient taste aversions to bitterness, sourness, and excess salt, eventually broadened to include noxious smells. But disgust is elastic. Darwin described it as "something revolting, primarily in relation to the sense of taste, as actually perceived or vividly imagined; and secondarily to anything which causes a similar feeling, through the sense of smell, touch, and even of eyesight." Almost anything, it seems, can provoke it: a touch, the sight of a sick person, gore, violence, a personal betrayal, sexual deviance, and classes of people. What do the senses of taste and smell have to do with this assortment of seemingly unrelated responses?

Basic tastes stir desires and gratification. Aromas summon memories and feelings. The brain effortlessly assembles these into sensations. Flavor is all in your head, a wholly internal experience. But *Homo sapiens* is an innately gregarious species that evolved living in groups, eating together, and cooperating to ward off danger. Human senses engage with the world—and other humans. Disgust is, in other words, a medium of communication. Its distinctive grimace is present from birth. "I never saw disgust more plainly expressed than on the face of one of my infants at the age of five months, when, for the first time, some cold water, and again a month afterwards, when a piece of ripe cherry was put into his mouth," Darwin wrote. "This was shown by the lips and whole mouth assuming a shape which allowed the contents to run or fall quickly out; the tongue being likewise protruded. These movements were accompanied by a little shudder." This is more than just a particular arrangement of facial muscles. It is a mediation between a person's own private universe of sensation and the life of the group, which lives or dies depending on its skill for communicating feelings and information.

Darwin researched faces with verve and invention. He asked scientists and missionaries around the world to gather evidence on the emotional responses of aboriginal peoples. He asked the young mothers he knew for anecdotes about the faces their children made. He collected friends' observations of their dogs. He commissioned or collected dozens of drawings and photos. This presented obstacles. Facial expressions, like the feelings they express, are fleeting, and photography techniques of that era required long exposure times. A subject would have to remain perfectly still, face frozen, for a minute or longer. Instead, Darwin obtained photos from the experiments of a French doctor, who had administered electricity to a patient who had lost all feeling in his face. This produced fixed expressions for as long as necessary, though the images had an unsettling appearance.

Emotions was wrong on some points. It argued a later-discredited concept that animals could inherit new facial expressions that their parents had learned. But over the past forty years, science has since shown that one of the book's basic insights was correct: facial expressions have biological and evolutionary roots.

In the late 1960s, the psychologist Paul Ekman visited members of the remote Fore tribe in the highlands of southeast New Guinea. He was testing an idea central to Darwin's book: because human facial expressions had evolved from those of animals, they transcended culture and conditioning and could be recognized anywhere on earth. Margaret Mead, the influential anthropologist, argued that culture was the force that molded human emotions and actions. A generation after World War II, suggesting human behavior was driven by biology or genetics was sometimes compared to eugenics, even Nazism. Darwin's book had been out of print for decades and almost forgotten, and its ideas had fallen into disrepute.

Darwin theorized there were six universal facial expressions, articulating happiness, sadness, anger, fear, surprise, and disgust. He believed that disgust, and perhaps happiness, were tied to food and flavor. Ekman had a million-dollar grant from the Defense Department to study facial expressions. He began his research with isolated Stone Age tribes. If their affects matched those of people in modern societies, that would demonstrate that the influence of culture had been overrated, and that there was something more elemental at work.

The Fore had already attracted scientific attention because they practiced ritual cannibalism, eating the brains of their dead. In the early 1960s this led to an epidemic of kuru, a disease that destroys brain tissue, producing tremors, seizures, dementia, and, ultimately, death. Both kuru and mad cow disease are caused by misfolded proteins called prions in brain tissue. By chance, Ekman found films of the Fore that National Institutes of Health researchers had made while studying the kuru epidemic.

Ekman spent months studying the faces in the movies. He observed how the Fore reacted to bad food, to pain, and to each other. "I found that Darwin was right," he said. "Because every expression you'd ever seen was in that culture. But the question was, how do you get scientific proof of this?"

He traveled the world, testing the reactions of college-age people in the United States, Japan, Brazil, Argentina, and

Chile. He found they could consistently identify the same basic expressions. But when he tested members of the Fore and one other tribe, the Sadong of Borneo, he found that their interpretation of some expressions did not match those of the college students. He wondered if his observations from the films had been off. But other factors might also be influencing the results. Working with Stone Age tribes had posed unusual obstacles. The tests required volunteers to read basic instructions and a list of emotions while responding to photographs of faces. But the Fore couldn't read, so a tester had to read instructions to them. It was also hard to translate words for specific emotions into their language. Ekman could also not be sure that the Fore hadn't picked up knowledge from the outside world that influenced their answers. Ultimately, he redid the tests with a twist. He recruited children who had had minimal contact with missionaries and other outsiders. Instead of a list of emotions, he used a set of very brief "stories" keyed to Fore culture, each of which captured a particular emotion. The story for disgust was "He/she is looking at something he/she dislikes," or "He/she is looking at something which smells bad."

These tests showed the Fore's facial expressions were nearly identical to those of people in developed countries like America or Japan. There were subtle differences, suggesting that cultural forces play a role in shaping these reactions: the Fore did not make the same distinctions between fear and surprise that other cultures did. While the Fore recognized disgust when they saw it in others, the things they found disgusting varied. But it appeared Darwin had been correct on another point as well: fundamentally, the differences between citizens of the British Empire and the inhabitants of Tierra del Fuego were not so vast after all.

• • •

The basic version of the "yuck" face sends a clear and highly useful alarm to others: *Spit that out!* Seeing it produces an empathetic wince. This form of messaging is indeed a legacy of human evolution. A lot of our formidable brainpower is devoted to making and understanding facial expressions. Humans, apes, and some monkeys have much larger primary visual cortices—the brain's initial processing area for sight—and larger knots of neurons devoted to the control of facial muscles than other mammals. These species live in larger groups than other primates do, with more complex social hierarchies. For early *Homo sapiens* groups, the rhythms of hunting, gathering, and preparing food, and then sharing and savoring it, would have encouraged ever more subtle and precise forms of communication. At some point, the spit-that-out wince of disgust, which many mammals display, began to serve new purposes. The most important of these was a warning against disease.

Disease poses a constant threat to groups. Unlike toxins in food, diseases have many different avenues of attack. Bacteria and viruses spread invisibly through food, physical contact, and insect bites. Early humans would have recognized the warning signs of possible infection: spoiled food, a festering wound, fever, a rash, vomiting. These would have evoked the earliest forms of a new, more expansive kind of distaste.

Valerie Curtis, a biologist at the London School of Hygiene & Tropical Medicine, devised a clever way to detect echoes of this ancient transformation amid the buzz of modern life. In 2003, she posted twenty photos of random people and objects to a BBC website. Visitors rated each photo's disgustingness on a scale of zero to five. Seeded among them were pairs of similar images, of which one had been altered to suggest disease. A picture showed a dish of blue liquid; its counterpart depicted what appeared to be pus and blood. Another photo showed a man's healthy face. In the altered version his skin was spotty, and he looked feverish. To suggest infection, Curtis included a photo of an empty subway car and one filled with people. Nearly forty thousand people around the world weighed in. Unsurprisingly, a majority found the disease-related images more disgusting, women more so than men; Curtis thinks such heightened sensitivity may have helped early human females protect babies and young children from sickness. A separate study by UCLA anthropologist Daniel Fessler found that women grow even more easily disgusted during the first trimester of a pregnancy, when their immune systems weaken to avoid attacking the fetus. When the risk of disease rises, the brain and body respond with heightened alertness.

As people age, their vigilance declines. The older the participants in Curtis's study were, the less offensive they found the disease photos. Curtis believes that this is because old people are less likely to reproduce, and so have less need, from the standpoint of natural selection and the group's survival, to watch for the warning signs of disease. Curtis also asked people to rank the person they'd least like to share a toothbrush with, from a list including "postal carrier," "boss," "TV weather forecaster," "sibling," "best friend," and "spouse." The more tenuous the bond, the more disgusting this idea was. Strangers pose a greater risk of disease to the unexposed immune system than friends or relatives do.

Curtis dubbed this suite of responses the "behavioral immune system." It's a set of cues blending the senses with group dynamics. These habits were built on observation, forbearance, and ultimately, success. Over the eons, the behavioral immune system would have constantly altered and expanded its contours to meet endless, changing threats. As the "yuck" face was applied to new things and situations, people would have combined it with language and gesture, creating an expanding expressive repertoire.

Distaste and the "yuck" face are the products of an ancient circuit of firing neurons, blood flow, and neurotransmitter activity in the brain that includes the insula and orbitofrontal cortex. Disgust uses the same circuit. But it has adapted this wiring for new purposes. A pleasant, easygoing man whom scientists dubbed Patient B. helped illuminate the inside of this black box.

In 1975, when Patient B. was forty-eight years old, he contracted a severe form of encephalitis, an inflammation of the brain caused by an infection of the herpes simplex virus. B. fell into a coma for three days, then awoke and gradually improved before being released from the hospital a month later. But B.'s brain, and his mind, were badly crippled. The infection had ravaged structures involved with memory and emotion, including the amygdala and the hippocampus of each hemisphere. He could remember events and dates from his childhood, but almost nothing later. He lived in a constant present, holding on to new facts for only forty seconds. His knowledge was mostly generalities: he couldn't recall his own wedding, but he knew what a wedding was. Nevertheless, those meeting him for the first time might not immediately detect a problem. He seemed happy. He laughed often, was an avid checkers player, and welcomed the neuroscientists who lined up to run tests on him. He enjoyed the mental challenges they provided.

Patient B.'s oddest quirks had to do with flavor. Parts of his insula and orbitofrontal cortex had been destroyed. He couldn't tell the difference between salt water and sugar water. He'd drink both with a smile, and would choose randomly if told to indicate the one he liked better. B. did have some taste perceptions, but they were mostly unconscious. In a 2005 experiment conducted by neuroscientists Ralph Adolphs and Antonio Damasio, B. was presented with salt water and sugar water, this time colored red or green. This changed everything. He was told to sample both and choose the one he liked; eighteen out of nineteen times he chose the sugar water. When asked to sip the saline solution, he vehemently refused. The colors created—or revealed—a preference for the sugar solution, without any awareness of or appreciation for the sweetness itself. Adolphs and Damasio theorized that there were undamaged parts of B.'s brain that could tell salty from sweet, but they were cut off from the damaged, conscious ones. Like a marooned man firing a flare gun to alert a passing ship, the colors allowed this part of the brain to signal its true feelings to the outside world.

B.'s sense of distaste was practically broken; so, unsurprisingly, was his sense of disgust. He had forgotten what disgust was, or even that it existed. He tossed back a cup of pure lime juice and pronounced it "delicious." When read a story about a person vomiting, B. said he imagined the person feeling hungry or delighted. Experimenters acted out facial expressions for him. B. recognized some of them, but identified disgust as "thirsty and hungry." When one of the researchers chewed some food and spit it out, making retching sounds and "yuck" faces, Patient B. again labeled the food "delicious."

Patient B.'s brain was too badly damaged to pinpoint

precisely where and how the disparate functions of feeling, imagining, and recognizing disgust came together, so neuroscientists embarked on a search. It led them to a familiar spot. In one experiment at France's National Centre for Scientific Research, fourteen volunteers had their brains scanned while they viewed movies of people reacting as they sniffed a glass containing a disgusting, pleasant, or neutral liquid. Then the fourteen were scanned as they did their own sniffing, and the results compared. The scans showed that observation and experience overlapped in only one spot: the anterior (forward) part of the insula, the area that processes tastes. It's also a place where inner feelings and outward, empathetic responses unite.

Feeling and observing disgust generate similar patterns of brain activity, and similar feelings. This is a basic form of empathy. Brain scans have shown that the more empathetic a person is, the more sensitive to disgust he is, and the brighter the insula burns. The insula, remember, is also a hub for many of the body's internal states and feelings. Its neurons align the taste system with brain structures that move facial muscles and recognize expressions, evoke memories, and enable speech, imagination, and storytelling. It also contains a distinct kind of neuron found only in the brains of humans, great apes, elephants, and whales and dolphins. Long, spindle-shaped von Economo neurons cluster mostly in the insula. They transmit messages across much longer distances than ordinary neurons, perhaps to bridge the ever-widening spans around the cortices of big-brained animals. Spindle neurons seem to help interpret and respond to emotional cues, shaping our relationships and social personae.

This means that visceral taste reactions underlie our most sophisticated behavior, animating our thoughts and judgments about everything from politics to money. Psychologist Hanah Chapman of the University of Toronto wanted to test this idea. She did an experiment in 2009 that focused on twinned muscles on either side of the mouth and upper lip that contract when a grimace is made, wrinkling the nose, called the levator labii. In the first phase, electrodes measured the muscles constricting in response to bitter drinks and photos of feces, injuries, and insects. Chapman then reran the experiment, this time with volunteers playing the Ultimatum game. Two players have a ten-dollar sum: one proposes how to split the money, the other decides to accept or reject the offer. If accepted, the money is split accordingly; if not, neither gets anything. Players rated their own emotional responses to the offers and outcomes as their facial muscles were monitored. As the offers grew more unfair, people became disgusted, their levator labii muscles twitched, and they were more likely to reject the offer. When their counterparts offered only one dollar out of the ten dollars, the contractions spiked.

The signal was clear: unfairness evoked the same muscular twitch as tasting something terrible. Rather than triggering anger, violating the everyday moral code of fairness led to revulsion, and to rejection of the unfair offers—and of the people making them. Taste had morphed into a primitive form of morality.

In the 1980s, Paul Rozin, a professor of psychology at the University of Pennsylvania, became fascinated with these gradations of disgust. At the time, no one else in his field was interested; the topic was considered marginal, a dead end. He decided to pursue it anyway. In a 1985 experiment, Rozin examined how the sense of contamination—the same feeling Darwin experienced when the Yahgan man touched the meat in his tin—emerges in children.

Rozin juxtaposed apple juice with a comb, and cookies with a dead grasshopper. Each pair of items was presented to a group of children aged three to twelve and a half, along with a scenario intended to provoke a particular degree of disgust.

First, a researcher told the children they could drink the juice after she stirred it with the comb. In one test, the comb was brand-new. In another, the volunteers were told it had been used but washed. In the third version, the story went, the comb had just been used on the researcher's own hair. Next, a dead grasshopper was placed next to a plate of short-bread cookies. The researcher sprinkled green sugar on the cookies, saying it was made from ground-up grasshoppers but tasted just like sugar. Finally, the experimenter poured some more juice, took out another dead grasshopper, and dropped it into the cup. It floated. She offered the child a straw and said, "Would you like to drink?"

The older the children, the more likely they were to reject the contaminated object. Eighty percent of the kids between three and six drank from the glass supposedly stirred with a used comb, but only ten percent of the eldest group. (Though a full 20 percent of this group decided to try the floating grasshopper–apple juice drink; perhaps adolescent daring had come into play.) When the same tests were run on adults, they were even more sensitive. Only five out of sixty-seven children refused the juice stirred with a new comb, while nearly half of adults did.

The sense of disgust evolves over the course of a lifetime. As children grow into adults, their social interactions become more complicated. At the same time, they pick up on social rules. Both of these imprint themselves on the brain. As people reach adulthood, the personal universe of disgust expands exponentially. Rozin and his colleague Jonathan Haidt divided it into four categories: inappropriate sexual acts, poor hygiene, death, and violations of body norms such as wounds, or obesity, or deformities.

Rozin saw an underlying theme. We hate to be reminded that we are all animals with fragile bodies that bleed, excrete, have sex, and get sick. These things remind us of death. We're the only animals who know death is coming; disgust is one method of averting our eyes from anything that reminds us of that.

Before humans can eat, animals and plants must die. The slaughter and rendering of animals is kept out of sight, and flank steaks, chicken parts, and spareribs appear disembodied behind glass and under plastic in supermarkets, as if by magic. In the United States, only certain kinds of animals are considered acceptable to eat: cattle, pigs, and fish, but not horses, dogs, or rats. Some organs—livers, calves' pancreases in sweetbreads—are delicacies. Others—bladders, hearts, and brains—are repulsive. These rules vary, seemingly arbitrarily, by location and culture. In the American South, pigs' feet are common fare; in Mexico, there's the offal in *menudo*, a tripe soup; in China, almost any part of a chicken is fair game.

In the extreme, entire nations can come to love things that repel outsiders. This makes the line between disgust and deliciousness razor-thin, moving depending on geography, climate, and culture. In Iceland, there's a popular dish called *hákarl*, made from the fermented flesh of the Greenland shark, which is notorious for its foul, ammonia-tainted flavor. *Hákarl* has become a hard-core foodie challenge, which even celebrity chefs routinely fail. Anthony Bourdain called it "the single worst, most disgusting and terrible-tasting thing" he had ever eaten. When Andrew Zimmern, host of the Travel Channel's *Bizarre Foods*, smelled it, he said he was reminded of "some of the most horrific things I've ever breathed in my life." But he at least found the taste tolerable. Gordon Ramsay spit it out.

Greenland shark meat is poisonous, a result of its peculiar physiology. Sharks filter out some wastes through muscles and skin rather than in their urine. The Greenland shark retains uncommonly high concentrations of urea, the main component of urine, and trimethylamine oxide (TMAO), a potent neurotoxin that induces a condition resembling extreme drunkenness and sometimes leads to death. During the late Middle Ages, Icelanders solved this problem by burying the meat in the sand, placing heavy stones on top to press the poisons out, and then leaving it for several months, just as the Yahgan did with their whale meat and blubber. As it froze, thawed, and refroze, *Lactobacillus* and *Acinetobacter* bacteria proliferated, producing enzymes that broke down the urea and TMAO. The smell, however, worsened because of two by-products. Urea decomposes into ammonia, and TMAO into trimethylamine, the compound that gives rotting fish its signature odor.

Bjarnarhöfn, a remote spot on the sea near a field of magma formations, is one of a handful of places in Iceland where *hákarl* is made. Kristján Hildibrandsson runs an operation that processes about a hundred shark carcasses a year, and a small museum dedicated to the tradition. Hildibrandsson's father and grandfather used to troll for sharks in a twenty-foot dory, but now they are purchased from giant trawlers at the dock. He compresses the meat in wooden crates for four to six weeks, then hangs and air-cures the mottled orange-yellow-gray bolts of flesh in a shed behind the museum for three to five months. A potent ammonia odor cloaks the shed for a radius of about fifty feet, even in the rain. Hundreds of years ago, semirotted shark meat may have been the only thing around in wintertime to sustain Viking colonies; today, it's eaten in bite-sized chunks with a snort of *brennivín* (Icelandic herbal schnapps), during Þorrablót, a midwinter festival dedicated to the Norse god Thor. Hildibrandsson invites visitors to sample small chunks of *hákarl*, accompanied by pieces of brown bread. "Some people like to have it right afterwards, to kill the taste," he said helpfully. The ammonia-rot aroma hits twice—once as the fish is brought out, and again when it's chewed—and overwhelms everything.

The rules that govern what is disgusting and what is a delicacy have no biological rationale; they are a product of complex societies that provide a variety of foods and have the luxury of drawing such lines based on tradition. "Almost all animal products are disgusting, and they are the most nutritive of all foods," Rozin said. "So why should you be so negative about something that is such a good package of nutrients and calories as meat is?"

The sources of disgust are endless. Rozin found that the fear of contamination is the most persistent. Once something is seen as contaminated, it may transfer that quality to everything it comes into contact with. The impurity may be metaphorical, but to the brain, it's quite real. Research shows that people with psychological conditions such as obsessive-compulsive disorder have an overdeveloped sense of disgust, leading them to take repetitive steps, such as hand washing, to dispel this sense of contamination. Having developed such an avid and promiscuous disgust sense, societies had to find ways to manage it. Hebrew laws regarding kosher food, for example, explicitly define what is contaminated and what isn't. In the Bible, God decrees that the Jews may eat only those land animals that have cloven hooves and chew cud, and only fish with scales; this excludes pigs, rabbits, and shellfish, among other things. Animals to be eaten must not be diseased, and must be ritually slaughtered by a single cut across the throat.

Disgust is also corruptible. It can become a cultural force that divides nations and peoples. "I believe disgust is an extremely dangerous emotion," Paul Ekman said. "It motivates genocide. When you believe people are repulsive, it dehumanizes them. Joseph Goebbels, Hitler's chief propagandist, wrote that the Jewish people were like lice, a disease. Those are disgust words."

Consider Charles Darwin's encounter with the Yahgan and the meat tin once again. Darwin was an archetypal humanist, seeking universal scientific explanations that transcended culture. But his reaction, and the tribesman's, reflected the impossibility of communication between them. Each lived in a different world, his perceptions of the meat tin defined by distinct childhood experiences and the rules of his respective society. Their reactions capture an inflection point in the long arc of flavor itself, as the rising modern world and its strange food inventions encountered the vanishing natural one that had forged human tastes.

To Darwin, the Yahgan man's touch was akin to contamination: whatever ineffable quality made the natives dirty was transferred to the meat. The fact that this was the touch of a human, and not an animal, made it worse. "I declare the thought," Darwin wrote in a letter to a colleague in 1862, "when I first saw in Tierra del Fuego a naked, painted, shivering, hideous savage, that my ancestors must have been somewhat similar beings, was at the time as revolting to me, nay, more revolting, than my present belief that an incomparably more remote ancestor was a hairy beast."

Darwin's reaction was a product of its time. The British Empire was in ascendance, and in once remote parts of the world men with backgrounds similar to his own were encountering tribal peoples and devising ways to subdue and "civilize" them. A parallel cultural obsession of the eighteenth and nineteenth centuries was "wild children" who had lived all or part of their lives apart from society, scrounging for survival. They lived astride the divide between nature and civilization, and traveled back and forth across it. Wild children, to put it in twenty-first-century terminology, had food issues. Lucien Malson, a French psychologist, gathered information on fifty-three feral children over six centuries, from 1344's "Hesse wolf-child" through 1961's "Teheran ape-child," identifying common themes in their stories.

Wild children usually lived off the land, eating nearly inedible foods. An Irish "sheep-boy" found in 1672 "was completely insensitive to the cold and would only touch grass and hay," Malson wrote. A girl discovered in 1717 in the woods outside Zwolle in the Netherlands had been kidnapped at sixteen months old and later abandoned. "She was dressed in sacking and living on a diet of leaves and grass." When they reentered society, their tastes appeared as alien as the Yahgans' had to Darwin. They rejected normal foods, devoured horrifying stuff, and didn't mind things other people find repellent, such as blood, feces, or filth. They had lived without human interaction, and thus had no behavioral immune systems or cultural cues to tell them how to react.

The wild boy of Aveyron was the most famous of these cases. Around 1800, Jean-Marc-Gaspard Itard, a young doctor, was working at the National Institution for Deaf-Mutes in Paris when a wild child arrived. Hunters had captured him naked in the woods in Lacaune, in the French Pyrenees, in

1797. He escaped, but was recaptured fifteen months later. The boy was initially diagnosed as an "idiot" with no chance of rejoining society. But Itard believed that wildness could be cured. He and the boy spent hours together as Itard attempted to socialize him. He named the boy Victor and kept a detailed record of everything they did. It was one of the earliest applications of the scientific method to psychology.

At first, the only thing that drew Victor's attention was food. He ignored all sounds except for the crack of a walnut being opened. He also ate acorns, potatoes, and raw chestnuts. Victor let Itard feed him warm milk and boiled potatoes. He spat out everything else. His senses were discombobulated. Sometimes he'd reach his hand into the boiling water for a potato without showing pain. Gradually, his sensibilities evolved. Within months he would eat nothing but cooked food. He took table manners to their logical extreme. Itard wrote: "The articles of food with which this child was fed, for a little time after his arrival at Paris, were shockingly disgusting. He trailed them about the room, and ate them out of his hands that were besmeared with filth. But at the period of which I am now speaking, he constantly threw away, in a pot, the contents of his plate, if any particle of dirt or dust had fallen upon it; and after he had broken his walnuts under his feet, he took pains to clean them in the nicest and most delicate manner." Victor's sense of disgust had become so robust, Itard wondered if he'd gone too far.

The tribesman squatting beside Darwin at the campfire, meanwhile, was repelled by the preserved meat's look and clammy feel. He couldn't tell for sure if it was food at all:

it bore only a passing resemblance to animal flesh, raw or cooked. At the time, many Europeans had never seen it either, and would have reacted the same way.

The meat tin was a new invention. In 1795, the French government, then a post-revolutionary council known as the Directory, had faced a problem. Its armies were fighting insurrectionists at home and, led by General Napoleon Bonaparte, foreign enemies in Italy and Austria. Smoking, curing, salting, and other ancient methods of preserving food had failed: convoys of rations spoiled, and entire armies starved. The Directory offered a reward to anyone who could devise a reliable method of preserving food. The effort would ultimately take fourteen years; the Directory would exist for only four.

Nicolas Appert, a forty-five-year-old confectioner and former revolutionary who had founded a Parisian sweet shop named Fame, accepted the challenge. Working with sugar, syrups, and preserved fruits (Appert also invented peppermint schnapps for use as an ice cream topping) had showed him that some foods could last indefinitely, depending on the method of preservation. He wondered if there was a single approach that would work on everything. A well-known method for preserving wines involved heat. He started there, experimenting with different bottles, jars, and tins. He found that if he placed food in a jar; sealed it with cork, wire, and wax to make it airtight; and then heated it in water for five hours, the food was edible weeks or months later.

This method killed the microbes that cause rot and prevented new ones from growing by cutting off the oxygen supply. Appert was unaware of this invisible process, but his approach obviously worked. Paris's chefs loved it. No longer slaves to the seasons, now they could have whatever they wanted year-round. "The peas above all are green, tender, and more flavorful than those eaten at the height of the season," one gastronome raved about the bottles in the Fame shopwindow. Appert packed up bottles of peas, partridges, and gravy and shipped them to the French military. Later, the navy field-tested the technique and embraced it. In 1810, then emperor Napoleon gave Appert the promised reward of 12,000 francs (about $32,000 today). Appert wrote a book, *The Art of Preserving All Kinds of Animal and Vegetable Substances for Several Years*, and opened a bottling factory. But bottles break. Peter Durand, an English businessman, won a patent in 1810 for a similar technique using iron cans covered with a layer of tin to prevent rust. A few years later, the British Navy adopted this method to preserve meat. By the time of the *Beagle*'s voyage, it was standard shipboard fare.

The rise of canning was part of a major shift in the world's eating habits and tastes. In a matter of decades, new technologies and farming techniques and the advent of railroads and steamships would make meat, especially beef, available to far more people than ever before. Scientists, meanwhile, began to use animal flesh as a template for experimentation in nutrition, form, and flavor.

Justus von Liebig, a brilliant chemist and contemporary of Darwin's born in Germany in 1803, helped bring this sea change about. Liebig performed a series of groundbreaking studies in organic chemistry, and invented nitrogen-based fertilizer after identifying the element as crucial to plant growth: agriculture was transformed. Next, Liebig turned his attention to food. His aim was to use science to manipulate nature, which he believed was highly inefficient at providing nutrition. Eventually, he hoped, new technologies would allow people to synthesize all the food they needed. He began to engineer food and formulate flavor based on scientific principles.

Liebig theorized that the juices contain a meat's most essential nutrients, and that searing is the best and only way to keep them from burning off. Hence the cook's nostrum that browning meat before cooking it seals in the juices. Liebig's idea contradicted centuries of kitchen practice—cooks tended to roast a meat some distance from the flames, then quickly brown the outside at the end—but by the mid-nineteenth century cooks were aggressively charring it instead. Liebig turned out to be wrong. Juice isn't that nutritious; brown meat too much and it quickly dries out. (Browning meat in moderation does make it taste better, releasing a wave of umami and Maillard chemicals; this is why it's still standard practice.)

Liebig's most important achievement in this area was the invention of a new kind of food. Even before he began his career, what Appert and other food preservers were doing had been not just unprecedented but fundamentally strange. Preservation involves stopping processes, such as fermentation, that create flavor, arresting the flow of time. Liebig went a step further: he made meat even more abstract, eliminating its pesky, frightening physicality. He boiled meat down to its essence, preserved it in a cube, and used the cubes to make broth that he believed could feed the world. Liebig's extract of meat, first developed in the 1850s and manufactured in a South American beef tallow plant, became a sensation. Liebig bouillon cubes are still manufactured in Britain. Flavor was not among their strengths; nor, despite Liebig's ambitions, was nutrition. But like many manufactured foods today that trace their lineage back to it, its uniform blandness was predictable and reliable.

As Darwin's Yahgan compatriot examined this strange, squishy substance, he could have no idea that he was seeing the future of food and flavor. The civilized world had judged the process of hunting, killing, cutting up, and eating animals, which had helped mold the human body and brain, and which his tribe still practiced, to be disgusting. These practices were a mark of savagery. Now technology had invented ways to make them virtually disappear. The less people knew about where food came from, the better.

CHAPTER 7

Quest for Fire

At the start of the twenty-first century, a coterie of amateur horticulturalists around the world began an unusual competition. Toiling in backyards, trading seeds, and seeking tips on the Internet, they pursued a goal that seemed more the domain of food science labs: cultivating the hottest chili pepper in the world. They were trying to dethrone the *Guinness Book of World Records* champion since 1994, the Red Savina, a smooth-skinned chili about the size of a Ping-Pong ball, and two hundred times as hot as a jalapeño.

For growing numbers of enthusiasts, sampling the super-hot burn of such chilies was both an exercise in culinary appreciation and a test of mettle. The gardeners believed that the potential of chili heat had barely been tapped. To unlock it, they cross-pollinated existing hot pepper plants or grafted one onto another, hoping to get offspring consistently hotter than either parent. To enhance pungency, some exposed their plants to heat lamps and underwatered them. Prospective record-setting chilies were sent to labs that assessed their concentrations of capsaicin, the chemical responsible for the burning sensation. The goal was to surpass the Red Savina's rating of 577,000 Scoville heat units, the scale that measures hotness.

There were setbacks. In 2006, the Red Savina, a strain of habanero, was finally beaten by farmers in India. Guinness named a new record-holder, the Bhut Jolokia, commonly known as the "ghost pepper" for its pale, milky color. It had grown widely for decades in northeastern India's Assam region. Its hotness hovered around 1 million Scoville units. But the hobbyists persisted, and soon they had a series of breakthroughs. In 2010 and 2011, the Guinness title changed hands three times in four months.

The first new record-holder was the Infinity Chili (1,087,286 Scoville units), bred in Lincolnshire, England, by a gardener named Nick Woods. It was quickly overtaken by the Naga Viper (1,359,000 Scoville units), grown by Gerald Fowler, a pub owner in Cumbria, England. "Hot enough to strip paint," he declared. Next was the Trinidad Scorpion "Butch T," grown by Australian planter Marcel de Wit (1,463,700 Scoville units). When de Wit took his first batch to Melbourne to make hot sauce, the cooks put on chemical protective gear to shield themselves from fumes and accidental splashes.

Meanwhile, a mortgage banker living in South Carolina named Ed Currie was also pursuing the Guinness record. He tended to hundreds of chili plants in greenhouses he had built in his yard from two-by-twos and white plastic sheeting. He had a growing hot sauce business, but craved the recognition and cachet of being a world record-holder. Currie believed he had a chili so hot it might hold the title for years. He named it Smokin' Ed's Carolina Reaper. It was a cultivar of the pepper species *Capsicum chinense*, known for its explosive heat; its wrinkled, blazing-red chilies were about an inch long and shaped like fists. A nearby university lab verified that Reapers scored consistently above 1.5 million Scoville units; some surpassed 2 million. Currie had submitted the paperwork to Guinness, but its world record verification process was known to take months, sometimes years. So he waited, and kept working to make his chilies even hotter.

Biologically, chili heat is neither a taste nor a smell, but a visceral, intrinsically unpleasant burning sensation. Animals hate it; humans embrace it with gusto. There are several unsatisfying scientific explanations for the chili's ubiquity in cuisine, and for why some people endure severe—though harmless—pain to savor the hottest. One theory is geographical: chilies are a common ingredient in the tropics, and eating them makes people sweat, which helps them stay cool. But this fails to account for the chili's expanding popularity in colder climates. Another theory is sensual: food science writer Harold McGee suggests that by inflaming nerves in the mouth and on the tongue, chilies make the palate temporarily more sensitive to touch and temperature, and flavors more vivid and pleasurable. But some scientific evidence shows that the burn actually obscures these sensations.

Chili heat makes no biological sense. It is a flavor koan. Sweet, bitter, sour, salty, and umami tastes predate humanity by hundreds of millions of years. But chili heat is a new sensation for *Homo sapiens*. The chili pepper evolved far from the East African Rift Valley where modern humans first appeared, in an area in the Andean highlands of South America spanning present-day Peru and Bolivia. Humans first tasted the chili only about twelve thousand years ago, while migrating south into the Americas from Asia; it has become a fixture in global cuisine in only the past five hundred years. Humans have always tested and embraced new flavors. The rise of chili heat shows that the flavor sense is expanding its range further still, incorporating new kinds of sensations. This has strange implications. The senses of taste and smell have deep connections to human physiology, playing roles in metabolism, emotion, and social interactions. What happens when a new type of flavor comes along, barraging brains and bodies with a powerful, mysterious neurochemical signal over hundreds or thousands of years? Like the flood of dietary sugar that it has paralleled in the past few centuries, the spread of chili heat is as much a massive experiment in human physiology and behavior as a culinary trend—except that chili may prove to be a boon, not a curse.

Like the bitterness of broccoli, the chili burn is, in essence, a weapon. As dinosaurs died out about sixty-five million years ago, flowering plants, then relatively obscure members of the plant kingdom, developed elaborate defenses to survive in a world of proliferating threats from the changing climate and newly dominant mammals. Roses developed thorns; chilies, capsaicin.

Chili peppers belong to *Solanaceae*, the nightshade family. Nightshade and mandrake are notorious for their toxins. Jimson weed produces hallucinations. Most domesticated *Solanaceae* plants, such as potatoes, tomatoes, and eggplant, had such noxious substances bred out of them in the past few thousand years. But a few, such as chilies and tobacco, are grown specifically to heighten the effects of their active ingredients, a class of chemicals called alkaloids. Besides capsaicin and nicotine, they also include caffeine and the active ingredients in heroin and cocaine. Alkaloids are also a fixture in the most vividly flavorful foods. Chocolate contains a suite of them, including phenethylamine, a mild amphetamine, and anandamide, the human neurotransmitter that stimulates the hedonic hotspots, prodding pleasure.

Why did chilies develop this powerful defense? The burning sensation deters animals. But some wild peppers are hot and others bland; if capsaicin were solely a mammal repellent, the bland ones would have no protection. Biologist Jonathan Tewksbury of the University of Washington studied this question, focusing on a wild species growing in the Bolivian highlands, *Capsicum chacoense*, that produces both hot and mild peppers. Bugs with pointed proboscises infect the chilies with a fungus that rots them and kills their seeds. Hiking through verdant Andean valleys, Tewksbury tasted peppers, studied their skin for bug puncture marks, and looked for signs of infection. He found the fungus was doing far more damage to bland chilies than to hot ones; capsaicin seemed to repel the bugs, kill the fungus, or both.

This did not explain why some chilies were bland, but as Tewksbury mapped the disease, another twist emerged. Bland chilies produced more, and hardier, seeds, and they predominated on higher, colder slopes. This might mean too much capsaicin somehow impaired their ability to reproduce, and that it was not essential at high altitudes, where the risk of fungus was low. The hottest chilies grew in warmer valleys. The maps also suggested that over thousands or millions of years, as chilies spread down mountainsides to lowlands, they grew hotter.

Birds, which cannot sense capsaicin, expanded the chili's range by eating the fruits and spreading the seeds in their droppings. By the time humans arrived, pepper plants were growing across South America, the Caribbean, and all the way to North America. People first tasted capsaicin's heat somewhere in Mexico. It was, almost certainly, a disappointing experience. But not for long.

Linda Perry, a paleobotanist at the Smithsonian Institution, was piecing together evidence collected from archaeo-logical sites around the Americas in 2005, looking for clues about prehistoric tastes, when she found something she could not explain. Perry's method is similar to Patrick McGovern's search for the chemical signatures of ancient beverages; she looks for evidence of the ancient meal. Many plants store car-bohydrates in microscopic ampules called starch grains. These are like fingerprints: they vary in size and shape depending on the plant that makes them. They pass through the human digestive tract and fossilize. To find them, paleobotanists painstakingly scrape detritus off tools and kitchen implements excavated from prehistoric homes. They pick apart fossilized feces. The different starch grains provide a vivid portrait of the meals, snacks, and diets of a particular place and time.

As Perry examined microscopic traces from sites across Latin America, a mysterious kind of starch grain popped up repeatedly alongside those from staples such as corn, pota-toes, and manioc root. This was puzzling, Perry thought; all the major starches in ancient American food had already been accounted for. Then she had a chance encounter.

"I was at a party, and they had these chili pepper hors d'oeuvre things, and a guy was explaining to me in rather gruesome detail that he couldn't eat them because they caused him distress," Perry said. "Probably not the best party in the world, but anyway. And I thought, that's strange, because these grains are usually left by undigested starch, and pep-per doesn't have starch. But maybe it does." She extricated herself from the conversation and returned to her lab. Some quick research revealed that chilies did indeed contain starch grains. Images of modern ones matched her ancient mystery starches.

Suddenly, the understanding of ancient American diets

changed. Before Perry's discovery, botanists believed that chilies had been cultivated in many places across the Americas. But there was little archaeological evidence, which usually rotted away. The biggest finds were caves in the highlands of central Mexico, where excavations of eight-thousand-year-old garbage heaps uncovered dozens of intact fossilized chilies. The evidence showed that people had first collected them wild, and had started to cultivate them six thousand years ago. The assortment included ancestors of today's jalapeño, ancho, serrano, and tabasco peppers. They also farmed maize, beans, squash, and avocado, all still used in Mexican cuisine.

Perry's starch-grain discovery proved that chili peppers had been in use across the Americas. They were an ancient craze to rival the modern one. Chili peppers were the best, most available source of spice to liven up diets heavy on bland, mushy maize, squash, and roots. The starch-grain traces stretched back six thousand years, to a village near the coast of Ecuador. Perry deduced that cooks there had chopped chilies up, pulverized them on grindstones, mixed this mash up in bowls and cooking vessels, and scattered the remains. About two thousand years later, rocoto peppers, a round red chili with a bite, were kept in the larder of a house two miles above sea level in the Peruvian Andes. Fishermen-farmers on San Salvador in the Bahamas apparently used a manioc grater to slice up chilies about a thousand years ago. And in coastal Venezuela sometime between AD 1000 and 1500, chilies were paired with ginger to liven up corn, arrowroot, and another tuber called *guapo*.

In 1492, Christopher Columbus arrived in the Caribbean on his first voyage to the New World. Making his way through the Bahamas to Cuba and then Hispaniola, he sampled the native cuisines, dining on yams and corn, a bread made from manioc, conches, and a six-foot iguana, which, he noted in his log, "tastes like chicken." The red, hot chilies the Indians mixed with sweet potatoes and corn caught Columbus's eye. After misidentifying the Caribbean as the Far East and the people living there as Indians, he added another misnomer to the list, calling the chili plant "pepper"—*pimiento* in Spanish—after black pepper, *pimienta*, an unrelated plant that produces a similar, though less powerful, sensation. Columbus believed it would make a profitable export. "There is also much *aji*, which is their pepper and is worth more than our pepper, no one eats without it because it is very healthy," he wrote in his log, using the Taino term for them. "Fifty caravels can be loaded each year with it."

But the chili pepper turned out to be virtually worthless to traders. The big-money spices of the day were relatively scarce: cloves and cinnamon grew only in the South Pacific; sugar production depended on mills and refineries. Chilies could be grown by anyone in a mildly warm climate; only seeds were required. They spread as a poor person's spice, traded hand to hand. Over a few decades—a blink of an eye from the perspective of human evolution, and even culinary history—chili heat blazed from one side of the world to the other.

The *Pinta* was likely carrying the first chili pepper seeds to reach Europe when it returned to Spain, landing at the port of Bayona on March 1, 1493. Word of the new spice spread quickly. Six months later, Pietro Martire d'Anghiera, a prominent Italian historian at the Spanish court in Barcelona, noted that Columbus had discovered a pepper "more pungent than that from the Caucasus."

Monasteries along the coasts of southern Europe collected many kinds of pepper seeds, and monks experimented, breeding both hotter and milder kinds. Hungarians embraced paprika as their national spice. In Germany, chilies appear in a 1543 guide to herbs by the professor of medicine Leonhard Fuchs, who carefully rendered them in woodblock prints (though he mistakenly thought they originated in India, calling them by the name "calicut" pepper, after Calcutta). Portuguese sailors used them to spice their food, and brought them to ports of call around the world. Chili peppers traveled to West Africa, and then to the Congo, by 1498. They appeared on the Chinese island of Macau, and inland in Szechuan. By 1542, three types of peppers were being cultivated in India. Curries, previously spiced with black pepper, suddenly flared with heat. Purandara Dasa, a composer of the time, wrote a song devoted to the red chili, calling it the "savior of the poor": "I saw you green, then turning redder as you ripened. Nice to look at and tasty in a dish, but too hot if an excess is used."

From south Asia, chilies spread to Siam and Burma, the Philippines, and beyond; they were already growing on some Pacific Islands when Europeans first arrived later in the sixteenth century. Chili routes soon began to double back on themselves: when Africans were captured and enslaved starting in the late 1500s, they brought food flavored with hot chilies back to the Americas.

Four hundred years later, chilies are found nearly everywhere in the world. Four thousand strains provide the sizzle for countless dishes, from Mexican *moles* to Thai curries. They're second only to salt on the list of most popular spices, outselling their next closest competitor, black pepper, by five to one. In the twenty-first century, a chili pepper renaissance has pushed heat to new levels. A generation ago, traditional varieties such as the Scotch bonnet and habanero peppers had set the upper bounds of hotness, clocking in at 200,000 and 300,000 Scoville units, respectively. Ghost peppers were considered too painful for Western palates. But tastes changed. Jalapeños and banana peppers became standard fare in the blandest salad bars. Reality TV shows followed hosts roaming the world, sampling outrageous dishes. The worldwide chili pepper trade is 25 times the size it was fifty years ago, the world's population only 2.2 times larger. Americans ate an average of three pounds of chilies a year in 1980. That number has more than doubled, and the upward arc continues.

The race to breed superhot peppers is the vanguard of this heat movement. Aficionados belong to a passionate subculture whose ethos falls somewhere between those of wine connoisseurs and *Star Trek* fans, obsessed with the minutiae of seed trading, cultivar purity, and the Scoville scale. The field tends to be male-dominated; studies have shown men favor chili heat more than women, and there is some competitive macho frisson in the experience—and the idea—of ultrahotness. "You've got to understand the chili-head mentality," said Jim Duffy, a chili grower who lives outside San Diego. "It's like the person who goes out shopping for knickknacks at garage sales: they've got to keep satisfying that thing. They throw so many plants in their backyard and their wife's going, 'What are you going to do with all those peppers? Where are we going to plant my cucumbers?' And they're like, 'Uh, I didn't think of that.' They went on my website, saw all the pretty pictures, the eye candy, just like a guy looks at the *Sports Illustrated* swimsuit issue."

There are thirty known pepper species, all of the genus

Capsicum (from the Greek *kapto*, meaning "bite"). Five have been domesticated. Ordinary bell peppers, which have no bite, are a variety of *Capsicum annuum*; in addition to the Carolina Reaper, *Capsicum chinense* varieties include habaneros and ghost peppers. When he began cultivating peppers in earnest in the early 2000s, Ed Currie gathered seeds from those and others from around the world—at first, the package deliveries, his greenhouses, and the odd smells emanating from his kitchen raised the eyebrows of neighbors, who called the police.

Every one of the two hundred kinds of peppers Currie grows rates above 200,000 Scoville units, the pungency of a habanero. But he paid attention to flavor as well. He was trying to build a hot sauce business, and hot sauces have many aromatic subtleties to bring out the flavors in foods. He coaxed notes of sweetness, chocolate, cinnamon, and citrus from his chilies. In his greenhouses, colors of these varieties explode: yellow, orange, white, purple, as tantalizing as jungle fruits would have been for our primate ancestors. The business grew slowly at first, then faster with the help of friends. An accounting firm agreed to do his taxes in exchange for a case of hot sauce. A next-door neighbor offered backyard space for a greenhouse. A friend lent him a few acres of fallow farmland south of town for growing space.

As his business grew, Currie experimented with several dozen superhot cultivars. The Carolina Reaper is the crowning achievement of an arduous process. It typically takes eight years to produce a horticulturally unique chili. Plants must be carefully segregated to prevent cross-pollination. Repeated crosses and certain traits must take hold, so that genes can be passed from generation to generation. Many growers try, but cannot produce consistently hot fruits, and abandon the effort. Currie claims to have accelerated this process by three years, and grown a strain that always produces a desired level of heat. At a university lab, samples are freeze-dried, ground into powder, and dissolved in alcohol: the resulting solutions are clear shades of red, yellow, and caramel. They are run through a gas chromatograph, which vaporizes them and measures the concentration of capsaicin to get the Scoville scale measurement. The Carolina Reaper averaged 1,569,700 Scoville units. Currie had some up-and-coming varieties that surpassed that figure. But Guinness was taking its time. "We've been going back and forth with Guinness for three years. I don't care if it takes another three years," he said. "Because what we're doing will last. World records can be one-off peppers that couldn't be reproduced."

The hucksterism surrounding the Guinness record led scientists at New Mexico State University's Chili Pepper Institute to take a more deliberate approach, growing top superhot varieties (though not Currie's) together under controlled conditions in 2011. The winner was the Trinidad Moruga Scorpion, another *Capsicum chinense* cultivar, ranked at 1.2 million Scoville heat units.

At a local restaurant, Currie opened a plastic zipper bag and dumped a pile of chilies, collected from his greenhouses, onto the table: Reapers, ghost peppers, and another variety called the Moruga Viper, in shades of red, orange, and bright yellow. He took a steak knife and carefully cut narrow slices of each, then passed the plate. The sensation of any superhot chili is not a raw blast; its qualities vary depending on the type of plant, the amount of capsaicin, and related substances called capsaicinoids, tempered by the complex chemical makeup of the pepper. Chili heat has three main elements. The first involves suspense: as chewing breaks open cell walls and releases capsaicin, there's a lag time between the initial bite and the perception of heat. This varies by pepper variety; habaneros have a particularly long delay of fifteen to twenty seconds. The second feature is dissipation. The heat from chilies in Thai cuisine tends to ease quickly, while varieties such as the ghost pepper linger. Finally, each burn has a unique feel. Asian peppers have a prickly heat; with American Southwest chilies, it's broad and flat.

I brought my teenaged son, Matthew, to taste the Carolina Reaper. He had been a dedicated heat freak since sometime not long after birth, preferring hot salsa on tortilla chips at the age of two. As he got older, he pursued this taste avidly, ordering the hottest dishes in restaurants, savoring habanero peppers as tears streamed down his cheeks. He seemed out to bludgeon his senses, reaching for a boundary between flavor and sensory overload, where taste blends into other visceral bodily responses and blots everything else out. The hottest known chili was the ultimate challenge, a Mount Everest of taste.

He put a small slice on his tongue. His hands went to his cheeks and mouth. He stood up, breathing hard. He ate some buttered bread; capsaicin dissolves in milk fat, which is why milk or butter helps cut the burn. He sopped up ranch dressing with another slice and chewed on that for a few minutes. He put a slice of lime in his mouth, hoping that the strong sensation would distract from the burn. I put a slice of Reaper, only a few millimeters square, on my tongue. The heat came in about fifteen seconds after a pleasant start, a taste of lemon and chocolate. The burning spread over my tongue and around my mouth and became overwhelming.

A kind of wave washed through my body. Voices around me receded as I slouched in my chair. I noticed my nose was running, and then I began to hiccup.

Currie hefted a whole pepper, bit off half, and chewed it nonchalantly. He sipped ice water (which has little effect on the pepper burn). His eyes watered a bit, but this soon passed. "I literally feel it in every inch of my body," he assured us. "But it is a positive feeling." Then he popped the rest in. We amateurs looked on, impressed, our mouths still burning.

Chili heat is painful, yet enjoyable; fiery, with no rise in temperature. In 1953, T. S. Lee, a biologist at the National University of Singapore, tried to unravel the physiology behind this reaction. He asked a group of forty-six young men to eat chilies, and monitored their sweating. Perspiration is a physiological reaction to heat. Rising body temperature, whether from the surroundings or from muscles warming during exercise, triggers a reaction in the hypothalamus. Via a series of feedbacks between the brain and the body, sweat glands go to work. Sweat evaporating off the skin cools the body; when its temperature drops back to normal, it stops.

Lee had the volunteers dress in cotton trousers only, then painted their faces, ears, necks, and upper bodies with a solution of iodine and dusted them with dry cornstarch—a combination that makes sweat turn blue. Lee used peppers common in Asian cuisine, from the species *Capsicum annuum*. Their tapered red fruits are about ten to twenty times hotter than jalapeños. For the sake of comparison, at a different time Lee's subjects also taste-tested solutions of cane sugar, bitter quinine, acetic acid, potassium alum (an astringent that makes the lips pucker), ground black pepper, mustard paste, and hot oatmeal. Some also gargled with hot water, chewed rubber, or swallowed feeding tubes.

In one experimental run, after eating chilies for five minutes straight, the subjects flushed red in the face, then all but one began to sweat. The areas around their noses and mouths turned blue, followed by their cheeks. Lee did another trial with seven participants, feeding them one pepper, then another: five continued to sweat, two profusely. Among the controls, only the acid and ground pepper made the volunteers sweat.

Eating chilies doesn't raise body temperature, so there is no physical need for cooling. Yet in Lee's experiment, the subjects sweated as if they had run a mile on a hot afternoon. To verify that the reactions to chili heat and genuine heat were equivalent, Lee had some volunteers put their legs in hot water. As their temperatures rose, the patterns of sweating on their faces were identical to those produced by eating peppers.

Lee had already deduced that chili heat could not be a taste, because people felt its burn on their lips, where there are no taste receptors. His experimental results indicated another body system was at work: the one that registers discomfort from burning. The chili burn was a form of pain. But it differs in one important respect: touch boiling water, and the pain intensifies until the hand is withdrawn. Start eating a Carolina Reaper, and the heat builds for several minutes, becoming overwhelming. But continue, and the heat recedes, leaving the mouth numb to chili's effects. Capsaicin causes pain, then blocks it.

Chili extracts have been used as painkillers for centuries or longer, stretching back into the pre-Columbian era. In 1552, a pair of Mexican natives, Martín de la Cruz, a healer, and Juan Badiano, a teacher, collaborated on a guide to Aztec herbal remedies now known as the *Badiano Codex*. It makes extensive use of the analgesic properties of chilies. One remedy for inflamed gums was to make a compress: boil the roots of several kinds of pepper plants along with a chili paste, wrap the resulting stew in cotton, and press it against the afflicted area. Elsewhere, native Americans rubbed hot peppers on their genitals to dull sensation and prolong their sexual pleasure—something early Spanish settlers also tried, to the dismay of prudish priests accompanying them. In nineteenth-century China, chili extracts were used as a local anesthetic for men about to be castrated to serve the emperor's court as eunuchs.

It was capsaicin's painkilling potential that the chemist Wilbur Scoville was trying to exploit when he developed his eponymous heat scale a century ago. He worked at the laboratory of one of the world's leading drug manufacturers, the Parke-Davis Company, outside Detroit. Parke-Davis and other pharmaceutical makers of the era were finding new ways to use plant alkaloids, including capsaicin and cocaine. (Parke-Davis once paid Sigmund Freud twenty-four dollars to rate its cocaine products, including a powder and an elixir, against those of its more established German rival, Merck. He noted only a small difference in taste, writing: "This is a beautiful white powder (available at a low price)."

Capsaicin was the active ingredient in Heet Liniment, Parke-Davis's topical painkiller cream. Scoville was assigned to measure the relative hotness of various pepper plants and concentrations of capsaicin, so that the correct dose could be more accurately gauged. Too much capsaicin burned unpleasantly; too little didn't work. Capsaicin had been isolated in 1846 by John Clough Thresh, who named it, and also noted that it was chemically related to vanilla. Capsaicin and its relatives, the most pungent compounds in the world, are molecular cousins to one of the gentlest, smoothest flavors. In 1912, there was no simple chemical test to detect capsaicin—only the sense of taste. Scoville ground up dried peppers and prepared extracts of different strengths. He assembled a panel of five lab colleagues. If a sample tasted hot, he diluted it repeatedly until no heat could be detected. The more dilution required to eliminate the last trace of burn, the hotter the pepper was.

Scoville had found a way to quantify a subjective sensation, an important achievement. He called it the Scoville Organoleptic Test, with heat measured in Scoville units. A rating of one million Scoville units meant that the extract had to be diluted to a concentration of one part per million before its heat disappeared. This approach was somewhat imprecise, because people have varying sensitivities to heat just as they do to other flavors, which is why today, the absolute concentration of capsaicin in a pepper is measured with a chromatograph and then converted to Scoville units.

Parke-Davis never succeeded in making capsaicin into an effective, profitable product. Heet is still sold today and still contains capsaicin, but the primary active ingredient is now methyl salicylate, derived from wintergreen. Today, five centuries after the *Badiano Codex* and one century after Scoville, drug companies are still trying to exploit capsaicin's numbing effect with dermal patches, injections, and other approaches, but success has largely eluded them. Manipulating the body's heat-sensing system is a dangerous business; in tests of one of these pain blockers, animals developed persistent high fevers: their bodies literally overheated.

Drug companies and biologists of Scoville's era who studied capsaicin's peculiar effects encountered the same obstacles that hampered the understanding of taste. They knew some kind of biological alchemy was occurring among capsaicin, body, and brain, but couldn't pinpoint how it worked.

Decades later, the key to this mystery was found in the milky sap of the resin spurge, a cactus-like plant that grows in the Atlas Mountains of Morocco. Moroccans slash open the plant, let the sap run out and dry, and harvest and sell the resulting resin, which contains the most powerful chemical irritant known: resiniferatoxin, or RTX for short, a form of supercapsaicin. Pure capsaicin rates at 16 million Scoville units; RTX rates at 16 *billion*, a thousand times hotter. Touching resin spurge sap causes severe chemical burns; swallowing more than an eyedropper full is fatal. Yet when greatly diluted, it has powerful medicinal qualities. In the first century AD, Juba, a North African king who was married to a daughter of Marc Antony and Cleopatra, had a terrible case of constipation, and his Greek physician Euphorbus prescribed some sap that had been dried, ground up, and cut with water. (The word "spurge" derives from the French word for "purge.") This ancient laxative worked so well that Juba named the plant "Euphorbia," after his doctor. Centuries later, Carl Linnaeus followed suit and named this genus of plants *Euphorbia*, and this particular one *Euphorbia resinifera*. Today, the resin is used to treat nasal blockages, snakebites, and poisons.

In the 1980s, RTX caught the attention of scientists studying the chili burn. Since it was so much more powerful than capsaicin, even the tiniest amounts made tissues flare in response. Research accelerated. When applied to the skin or injected, scientists found that RTX tricked the brain and body into thinking that room temperature was hotter than brimstone; then it abruptly shut down the body's ability to sense heat, or respond to any temperature changes. Rats treated with RTX developed hypothermia. But unlike a topical anesthetic, which numbs all feeling, RTX did not impair other kinds of touch; the rats could still feel a pinch or an electric shock. Only nerves that sensed heat were affected. In one experiment, scientists irradiated a bit of RTX to make it traceable by scanner, injected it into a cell, and observed as its molecules attached themselves to a previously unknown kind of receptor: a heat receptor.

Lee's sweat-while-you-eat experiment from forty years earlier had been vindicated. Both RTX and capsaicin molecules attached themselves to the body's receptors for registering heat and pain. These are part of a larger family of sensors that detect grave threats: heat, cold, burns, blows, cuts, pinches, electrical shocks. Without them, humans would die rather quickly.

Capsaicin receptors are embedded in the surface of nerve cells in the mouth, skin, eyes, ears, and nose. When these cells come in contact with anything hotter than 108 degrees Fahrenheit—the signal that the line between "toasty" and "too hot" has been crossed—the receptors' shape changes in response. This opens a pore to the cell's interior. The water in the human body is a salty soup of positively and negatively charged particles, diffusing in and out of cells. The pore is only one or two atoms wide, and allows only positively charged calcium ions through. The electrical charge makes the neuron fire, sending a signal to the brain. This process takes only milliseconds, a much faster reaction time than that of taste receptors. Thus the hand jerks away from a hot pan before awareness catches up.

Chilies trick this system. Begin eating a hot pepper, and

capsaicin molecules inundate these receptors. This lowers the mouth's heat threshold, much as salt lowers the melting point of ice; suddenly 98.6 degrees feels like 150. This is why chili peppers taste hot. The heat alarm reaches the brain via the trigeminal nerve, one of the major neural pathways in the head, relaying signals from the face, the nose and mouth, and the eye. The chili burn is the strongest of a number of "tastes" sensed by receptors for heat or touch and carried by the trigeminal nerve that include the sharp pungency of horseradish and wasabi, the gentler tang of lemongrass, and the hot tingle produced by the Szechuan pepper (which is not related to chili or black pepper). Szechuan peppers are also used as an ingredient in lipsticks to inflame the skin and simulate the sensation of pouty lips.

So, pain is a distinct part of flavor, with its own unusual properties. Heat receptors are present all over the body, which makes superhot chilies dangerous. An unpleasant taste can only be sensed by the tongue, but capsaicin envelops you, as my son and I found while watching Ed Currie prepare hot sauce. He poured a bottle of white Bhut Jolokia pepper mash—a beige-colored, six-to-one mix of pureed pepper and vinegar—into a pan, blended in additional spice, then put the mix on the stove. Capsaicin in the steam stung our eyes, then reached our noses, and we coughed and sneezed for ten minutes. Currie appeared immune.

Pepper spray, usually made from cayenne chilies, works the same way. Police-grade pepper spray rates above 5 million Scoville units, more than enough to cause temporary blindness, constricted breathing, near-total incapacitation, and in rare cases, death. India has led the world in finding ways to exploit these properties. The Indian army has experimented with ghost-pepper grenades, and with a food supplement to help warm the bodies of soldiers in the Himalayas. Adopting a local practice of farmers, the environmental agency in Assam set up fences made with ropes dipped in ghost-pepper oil to keep roving elephants out of agricultural areas. Elephant hides are too tough for electric fences, but they yield to ghost peppers.

Capsaicin affects the inside of the body, too. Like taste and smell receptors, heat receptors have been found in nerve cells nearly everywhere, including the brain, bladder, urethra, nasal membranes, and colon. Exactly what they all do isn't clear, but it goes beyond regulating the temperature; some help keep metabolic systems running within certain limits. But they may also be a source of serious health problems. In 2014, researchers led by Andrew Dillin of the University of California, Berkeley, ran an experiment with mice that had been genetically engineered to lack capsaicin receptors. Predictably, the mice had impaired reactions to heat. But they also lived four months, or 14 percent, longer than did normal mice, and had more youthful metabolisms. As normal mice aged, Dillin found the capsaicin receptors in their bodies started to malfunction. In some mice, they stimulated the pancreas to release a protein that made blood sugar chronically higher—a common malady of old age, and a precursor to diabetes.

Of course, people hoping to live longer can't get rid of their capsaicin receptors. But eating chilies does the next best thing: it paralyzes them. The numbing from eating superhot peppers occurs as receptors become overwhelmed and nerve cells shut down. The neurons usually recover, but sometimes they die. Julia Child once claimed that eating too much spicy food destroyed the taste buds; this is not true, but she was onto something. Inside the body, this blocking action may shut down the malfunctioning receptors, mimicking the effects that helped those mice live so long.

Many studies have shown that eating a chili-rich diet has small but measurable health benefits. Capsaicin raises metabolic rates, burning more calories. Mice without capsaicin receptors, with their active metabolisms, were also less likely to get fat. (Currie lost 180 pounds on a diet heavily spiced with superhot peppers, and says they helped him give up alcohol.)

No health benefit explains the other great mystery of chili heat: why people *enjoy* the pain and irritation. Like the affection for a hint of bitterness in cuisine, it's the result of conditioning. But there are no contests to brew the world's bitterest coffee. The chili sensation mimics that of physical heat, which has been a constant element of flavor since the invention of the cooking fire a million or more years ago; we evolved liking hot food. The chili sensation also resembles that of cold, which is also unpleasant to the skin, but pleasurable in drinks and ice cream, probably because we've developed an association between cooling off and the slaking of thirst. But none of these examples explains why, when nature devised capsaicin to repel, humans embraced it in spite of itself.

Paul Rozin became interested in this question in the 1970s, when his then wife Elisabeth Rozin wrote *The Flavor-Principle Cookbook.* Its theme was that ethnic cuisines had certain common flavor signatures that home chefs could appropriate to enliven meals. He began to wonder why some cultures favored highly spicy foods, and others bland. He traveled to a village in the highlands of Oaxaca, in southern

Mexico, to investigate, focusing on the differences between humans and animals. The Zapotec residents there ate a diet heavy in chili-spiced food; Rozin wondered if their pigs and dogs had also picked up a taste for it. "I asked people in the village if they knew of any animals that liked hot pepper," Rozin said. "They thought that was hilariously funny. They said: no animals like hot pepper!" He tested that observation, giving pigs and dogs there a choice between an unspicy cheese cracker and one laced with hot sauce. They'd eat both snacks, but always chose the mild cracker first.

Next, Rozin tried to condition rats to like chilies. If he could get them to choose spicy snacks over bland, it would show that the presence of heat in cuisine was probably a straightforward matter of adaptation; animals—and humans—liked heat because chilies were nutritious and the imperatives of survival had overcome its bad taste. Humans might have gradually grown less sensitive to it, just as the Aymara of Bolivia became accustomed to high levels of bitterness in their potatoes.

He fed one group of rats a peppery diet from birth; another had chili gradually added to their meals. Both groups continued to prefer nonspicy food. In another attempt, he spiked pepper-free food with a compound to make the rats sick, so that they would later find it disgusting. They still chose it over the chili-laced food. He induced a vitamin B deficiency in some rats, causing various heart, lung, and muscular problems, then nursed them back to health with chili-flavored food: this reduced but didn't eliminate their aversion to heat. In all, Rozin thinks he may have made only a single one of these rats into a chili convert. Only rats whose capsaicin-sensing ability had been destroyed truly lost their aversion to it. Rozin's only real success training animals as chili lovers came later, when he coaxed a pair of chimpanzees to develop a taste for chili-spiced crackers.

Rozin came to believe that something unique to humanity, some hidden dynamic in culture or psychology, was responsible for our love of chili's burn. For some reason apparently unrelated to survival, humans condition themselves to make an aversion gratifying. The Zapotec weren't born liking chilies, but picked up a taste for them around the age of four to six years old.

Not long after, Rozin compared the tolerances of a group of people in the United States with limited heat in their diets to the Mexican villagers' tastes. He fed each group corn snacks flavored with differing amounts of chili pepper, asking them to rank when the taste became optimal and when it became unbearable. Predictably, the Mexicans tolerated heat better than the Americans. But one thing was consistent for both groups: the difference between "just right" and "ouch" was razor-thin. "The hotness level they liked the most was just below the level of unbearable pain," Rozin said. "So that led me to think that the pain itself was involved: they were pushing the limits, and that was part of the phenomenon."

The chili culture is all about pushing limits. Ed Currie believed embracing it had helped him overcome his own weaknesses. He had organized his life around a single, powerful sensation, and it had worked: Guinness named Smokin' Ed's Carolina Reaper the world's hottest pepper in 2013. But success depended on staying ahead of the competition; the race would eventually take chili heat higher and higher, past two million Scoville units, into realms of pungency never tasted before. How far could he go, and who would follow?

Pleasure is never very far from aversion; this is a feature of our anatomy and behavior. In the brain, the two closely overlap. They both rely on nerves in the brainstem, indicating their ancient origins as reflexes. They both tap into the brain's system of dopamine neurons, which shapes motivation. They activate similar higher-level cortical areas that influence perceptions and consciousness. Anatomy suggests these two systems interact closely: in several brain structures, neurons responding to pain and pleasure lie close together, forming gradients from positive to negative. A lot of this cross talk takes place in the vicinity of the hedonic hotspots—areas that bridge basic reflexes and consciousness.

In behavior, pleasure and aversion also work in parallel. Each is a form of motivation forged by natural selection; each triggers actions to safeguard immediate survival and guide learning for the future. Pain alerts people to stop, to pull away, to avoid. Pleasure is a green light to continue, and to return. A little pleasure can reduce pain, and pain can temper pleasure; chronic pain can lead to depression and an inability to experience pleasure. Humans routinely endure pain to achieve a greater reward and the pleasures that accompany it; childbirth, for example. The opposite happens as well, when pleasure leads to pain, such as a hangover, or years of indulgence in drugs makes life seem meaningless and depressing.

The love of heat was nothing more than these two systems working together, Rozin concluded. Superhot tasters court danger and pain without risk, then feel relief when it ends. "People also come to like the fear and arousal produced by rides on roller coasters, parachute jumping, or horror movies," he wrote. "They enjoy crying at sad movies, and some come to enjoy the initial pain of stepping into a very hot bath or the shock of jumping into cold water. These 'benignly masochistic' activities, along with chili preference, seem to be uniquely human." Eating hot peppers may literally be a form of masochism, a soliciting of dangers that civilization cocoons us against.

Rozin's theory suggests that flavor has an unexpected emotional component: relief. A study led by Siri Leknes, at Oxford University, looked at the relationship of pleasure and relief to see if they were, in essence, the same. Leknes gave eighteen volunteers two tasks while their brains were scanned: one pleasant, one unpleasant. In the first, they were asked to imagine a series of pleasurable experiences, including consuming their favorite meal, drink, or cup of coffee or tea; the smell of a fresh sea breeze or freshly baked bread; a warm bath or smiling faces. In the other, they were given a visual signal that pain was coming, followed by a five-second burst of 120-degree heat from a device attached to their left arms: enough to be quite painful, but not enough to cause a burn.

The scans showed that relief and pleasure were interwined, overlapping in one area of the frontal cortex where perceptions and judgments form, and in another near the hedonic hotspots. As emotions, their intensity depended on many factors, including one's attitude toward life. Volunteers who scored higher on a pessimism scale got a stronger surge of relief than did optimists—perhaps because they weren't expecting the pain to end.

Ed Currie's website features videos of people eating Carolina Reapers. They are studies in torture. As one man tries a bite, his eyes open with surprise, then his chair tips back and he falls on the floor. Another sweats up a storm and appears to be suffering terribly, but presses on until he has eaten the whole thing. Watching these, it suddenly seemed clear to my son and me that whatever enjoyment might be derived from savoring chili flavors, true satisfaction comes only in the aftermath: the relief at having endured, and survived.

CHAPTER 8

The Great
Bombardment

During World War II, Ireland endured a regime of meager rations, bland flavors, and overall deprivation called the Emergency. With most resources directed to the war effort, housewives stood in line with coupon books to obtain basic items including tea, sugar, butter, flour, and bread. People were allowed to cook with gas for only a few hours per day, while utility company agents nicknamed "glimmer men" went door-to-door to ensure pilot lights were otherwise kept off. But a thirty-year-old entrepreneur named Joe Murphy saw opportunity in scarcity. The Irish craved fresh fruits, the only source of essential nutrients including vitamins C and D; imports had been cut off as German U-boats patrolled the Atlantic. In Britain, people satisfied this need with a drink named Ribena, a cordial made from the syrup of black currant berries. The British government had mandated the cultivation of the berries early in the war, and distributed Ribena to children as a nutritional supplement. Murphy secured a supply for Ireland: every bottle sold.

After the war ended, basic items such as meat, butter, and cheese became available again, and a modest array of snack foods helped fuel the baby boom. Soft drinks were stocked on shelves next to bottles of Ribena. In 1954, Murphy founded a potato chip business. Potatoes were Ireland's national staple, used as the foundation of stews and shepherd's pies; boxty, a potato pancake; and champ, mashed potatoes with scallions and milk. Yet in the 1950s, Ireland still imported all its potato chips from the United Kingdom. Investing five hundred British pounds, Murphy began with two rooms, two deep fryers, one van, and eight employees. He named his company "Tayto," after his toddler son Joseph's mispronunciation of the word.

Though his chips sold well, Murphy was not satisfied. During the war, the challenge had been finding products to meet demand. Now he faced a problem of oversupply. Large snack brands dominated the market, and with a product that was identical to theirs, his business seemed unlikely to grow. Dissatisfied with what he called the "insipid" flavor of plain chips, Murphy decided to give the senses a rattle. He began sifting cheese or onion powder on the potatoes as they came out of the fryer and selling the spiced chips as a novelty. One day when they had an excess of both ingredients, Murphy's partner, Seamus Burke, sat down at a table and combined them into a third flavor. These were the first flavored potato chips.

In the 1950s, American potato chip manufacturers faced the same monotony problem on a much larger scale. In Ireland, the cheese-and-onion combination quickly caught on, and Murphy's business grew rapidly. So American companies appropriated the idea and began making their own flavored chips. In 1958, Herr's Potato Chips of Lancaster, Pennsylvania, introduced barbecue chips, as did H. W. Lay & Company around the same time. More followed suit. Then came sour cream and onion, the American riff on cheese and onion.

The spiced potato chip was one of the first modern junk foods. As companies continued to experiment with seasonings, textures, and chemical formulas, the chip was transformed. It became a kind of industrial flavor template, generating profits by delivering precision-engineered jolts of flavor to the senses. Sixty years later, the diversity of potato chips is astonishing. Many flavors are lifted from world cuisines and tailored to suit local tastes. There are hot chili squid chips in Thailand, red caviar chips in Russia, shrimp and garlic chips in Spain, a chip in Australia flavored with Vegemite. In Britain, there's a chip that tastes like Yorkshire pudding.

The potato chip's metamorphosis was part of a contemporary sea change in food and tastes, comparable to controlling fire or fermentation. From the time people started cultivating grains twelve thousand years ago until about 1900, most of humanity lived on limited, starchy diets. The foundation of the meal was harvested grains or roots such as potatoes. Meats, milk, eggs, fruits, or vegetables were luxuries, available only occasionally. Food historian Rachel Laudan calls this "humble cuisine." Most of its calories came from a basic starch such as millet or maize; women prepared it and families ate out of a common bowl, sometimes using their fingers.

The wealthy ate quite differently. They built elaborate kitchens, employed teams of chefs, purchased animals for slaughter, and obtained foreign spices. Their food—high cuisine—was a diverse and sumptuous affair, embodying power and status. Meats, sweets, fats, and alcoholic beverages provided most of the calories. There were condiments, prepared sauces, multiple courses, and a sense of tradition and ritual. When starches were employed, it was the costlier grains such as rice and wheat.

By the twentieth century, sprawling, interlocking industrial food systems began doing what palace cooks and personal chefs had always done. They raised, killed, and rendered cattle, manufactured cheese and beer, grew and processed wheat, and formulated the recipes for condiments. Millions, then billions, of people around the world could now taste the seared meats, varied courses, and desserts once reserved for princes. A cheeseburger and french fries might sit at the farthest imaginable point from "haute" on the culinary scale, but it's a distillation of five thousand years of high cuisine. Laudan calls this "middling cuisine." Savory beef, salty-fatty fries, spiced-up ketchup and mustard, umami-rich cheese, and piquant onions are a court feast in microcosm. In the early 1900s, a period of food and flavor democracy began, bringing more robust nutrition to the masses.

But gradually, this system went haywire. As chips and other tasty snack foods proliferated, food companies competed to see how many sensory buttons they could push at once. Examining emerging insights into the biology of flavor, they found new ways to manipulate perceptions and desires, from genetics to cognitive tricks. Processed foods—sold in supermarkets or fast-food chains—bombarded the senses and toyed with the brain and the gut. As consumers became inured to these, food companies ladled on even more sensations. This era of overstimulating junk food dazzled the palate, but it harmed public health. In the United States, a Harvard study that tracked the diets and weight of 120,000 healthy men and women over two decades showed each gained an average of nearly a pound every year. In descending order, the extra weight was tied to potato chips, potatoes, sugary beverages, and red meat. Supertasty potato chips had become a leading cause of death.

• • •

The potato was once the quintessential humble cuisine. Like the chili pepper, it is descended from wild ancestors in the Andean highlands (and also belongs to the nightshade family, *Solanaceae*). Wild potatoes are knobby and bitter, yet early Americans recognized they packed a lot of nutrition into a small space and began to cultivate them. The Incas developed sophisticated systems for growing, storing, and preserving their potatoes. They used a series of steps to neutralize the roots' bitterness, which is caused by a pair of alkaloids, solanine and tomatine. Frozen at night and dried in the sun, spuds were trampled and then soaked to remove the skins and soften the flesh, and finally sun-dried one last time. The resulting cakey substance, *chuño*, could be stored for months and transported easily. It's still made today. Chilies were often used to flavor it.

When the Spanish brought potatoes to Europe in the 1500s following the conquest of the Inca, their bitter taste put people off. But the nutritional advantages ultimately proved too great to resist. Facing wars and revolution in the eighteenth century, the French embraced the potato after a pharmacist named Antoine-Augustin Parmentier subsisted on them during several stints as a prisoner held by the Prussians in Germany during the Seven Years War. When freed, he promoted the potato as the ideal solution to Europe's repeated famines. Marie Antoinette appeared in public wearing a garland of potato flowers in her hair; her husband, King Louis XVI, wore them as boutonnieres. The potato fueled a century of European population growth—though over-reliance on it brought disaster to Ireland in the nineteenth-century potato famine.

The potato chip is an American invention. Its exact origins are cloudy; the most famous story suggests it was born at the Moon Lake Lodge in Saratoga Springs, New York, in 1853, when a diner found his side dish of fried potatoes soggy. He sent them back to the kitchen. The chef, George Crum, made more, but the dish was rejected again. Crum did not appreciate being second-guessed. He thinly sliced some potatoes, threw them in the frier, and sent the browned, crisped, salted chips back out to his guest: the culinary equivalent of sarcasm. The dish was also delicious, and word spread. Soon chips were sold everywhere, from barrels in grocery stores and from horse-drawn carts. By the early twentieth century potato chip factories dotted the eastern United States. Most of them had begun as mom-and-pop operations like Joe Murphy's Irish one, hatched in somebody's back room, garage, or barn.

Combining the bland, white potato with boiling oil and salt had turned it into something quite different. Potato chips are innately gratifying knots of energy. Starches, fats, and salt have a synergy more powerful than that of umami: they give a powerful rush to the brain's pleasure centers, sparking delight and craving in equal measure. Their satisfying crunch is a learned, Pavlovian signal of freshness and imminent deliciousness. Bite into a chip, and the brain instantly recognizes something good is at hand. This snap judgment draws on tastes and aromas. But there are more forms of flavor perception involved that science has only just begun to decode. The emerging understanding of them may require revisions to the concept of basic tastes that dates back to ancient Greece.

Though starches are tasteless, the mouth may be able to sense them unconsciously and alert the brain. A 2014 fMRI study done by scientists at the University of Auckland, in New Zealand, found that volunteers who rinsed their mouths with a starch solution received a 30 percent boost in activity in the visual and motor cortices over a group that rinsed with a control solution. In response to this signal of food energy, the volunteers' attention became more focused and acute. How the mouth detects it was not clear. But the tongue clearly has capabilities besides the five tastes.

Scientists once attributed the powerful allure of fatty foods to their creaminess and rich aromas, but recent research suggests that fat is actually a sixth basic taste: there are fat receptors on the tongue that trigger a unique and pleasing perception. This makes sense. Like starches and sugars, fats are essential nutrients. They metabolize into fatty acids, a major energy source for cells. When human ancestors began to eat meat, the dietary surge in fatty acids helped make brains grow larger. A mutation unique to humans assisted in this transformation by helping to burn cholesterol and cutting the risk of heart disease from fat—at least, until people began to consume so much fatty food that this advantage was neutralized.

The fat taste is a molecular waltz: one type of receptor protein acts as a kind of chaperone for fat molecules, helping them to bond with a second receptor that alerts the brain. The more of these chaperone-proteins there are on the tongue, the greater a person's sensitivity to fat, and the richer it tastes. Some people are thousands of times more sensitive to fats than others. The fat-insensitive are more likely to be obese; according to one theory, they get little pleasure from fats because they can barely detect them, and overeat to compensate. They crave fat because they cannot get enough of it. But the overstimulation only dulls the fat sense more, a vicious cycle similar to drug addiction and sugar overindulgence.

Then there's salt. All life came from the sea, and some

400 million years after the first land-based animals appeared, the sea remains with us. The nervous system's messages travel via the electrically charged salt ions in the body's tissues. Healthy blood plasma and hydration both depend on keeping up a certain salt concentration. The body fastidiously husbands its salt to maintain this balance. An overdose can fatally disrupt it. As salt rises in the bloodstream, water diffuses from tissues into the blood to equalize the concentrations. An overwhelming thirst descends, muscles weaken, and the brain shrinks. Salt deprivation also kills. The body can manage for weeks without consuming it, but when the internal supply nears total exhaustion, an insatiable craving called sodium hunger kicks in. This is another example of the amazing plasticity of taste: in dire straits, seawater becomes delicious (and drinking it in this state is not fatal). When salt-starved rats drank a solution three times as salty as seawater in Kent Berridge's lab, neurons in their hedonic hotspots fired in the same pattern as they did when the rats drank sugar water: they loved it.

The salt taste—good in small amounts, awful in large ones—is calibrated to keep the body safely between those extremes. Neuroscientist Charles Zuker of Columbia University unraveled the mystery of how these contradictory sensations work. A salt receptor detects the "just right" taste of a pinch. But a sufficient concentration will activate bitter and sour receptors, triggering a highly unpleasant fusion of these two bad tastes.

This dual identity is a fixture of both cuisine and culture. Salt's talent for enhancing flavors is unmatched, and it has been used since prehistoric times as a preservative, flavoring, and all-purpose utility ingredient. It counters bitterness, makes fats tasty, makes soups and other liquids seem more robust, and boosts overall pleasure. A pinch of salt on baking bread catalyzes Maillard reactions, turning it golden brown. Prehistoric herders managed the movements of their livestock by handing out dabs of it, guiding animals to salt licks to keep them healthy. Sauces, salad, and sausages all derive their names from the Latin *salsus*, or "salted." "You are the salt of the earth," Jesus declares in the Gospel of Mark, comparing its presence to the teachings of God. An excess of salt, meanwhile, is a common symbol of barrenness and death; remember what happened to Lot's wife in the Book of Genesis. Fleeing the city of Sodom before its destruction, she ignored divine instructions telling all to not look back, and was turned into a pillar of salt.

Like fat, sugar, and starches, salt has become a story of excess. People in developed countries consume ten times more salt than remote tribesmen—and our salt-poor hunter-gather ancestors—raising the risk of cardiovascular disease. Yet food today doesn't taste supersalty—people are conditioned not to notice. This was once a survival strategy: our ancestors would overindulge whenever possible, so they could hold out should salt become scarce. Now it's killing people. Alan Kim Johnson, a behavioral psychologist at the University of Iowa, has studied these cravings and concluded that most of the world is literally addicted to salt.

The body's avid responses to carbohydrates, fat, and salt should have made the plain potato chip the ideal consumer product, or even the only food anyone could ever want. But this appeal has certain biological limits. Humans are omnivores, descended from hunter-gatherers who were in turn descended from scavenging mammals. The taste for novelty, variety, and contrast in food is a powerful and underappreciated force.

French psychologist Jacques Le Magnen discovered its

basic dynamic in the 1950s while studying the nature of hunger. Left blind at thirteen by a bout of encephalitis, Le Magnen compensated with a prodigious memory for scientific facts and data. He made his reputation studying the strange attractive powers of smell: how a nose's sensitivity may vary depending on signals from sex hormones, or just the time of day. Le Magnen then began examining the rhythms of appetite and feeding. He set up equipment to track every drop of drink and crumb of food a rat consumed in the course of a day. Almost immediately, he noticed something odd. Rats fed a single kind of food always stopped eating after a short while. Those who received an assortment of items continued indefinitely—and, over time, they gained weight. Some biological quirk was nudging the rats to eat a more diverse diet. Satisfying hunger was not a straightforward matter of consuming sufficient calories.

Human appetites function a lot like those of Le Magnen's rats. Eating any single food, even a delicious one, quickly becomes boring, then intolerable. Some American prisons exploit this effect to punish unruly inmates, serving them a blend of leftovers and basic food items such as rice, potatoes, oatmeal, beans, and carrots baked into a bland, indeterminate gray mass called "the loaf." It meets all basic nutritional needs, but prisoners universally despise it. The American Civil Liberties Union calls this taste deprivation a form of cruel and unusual punishment, but it seems to work: in one jail in Wisconsin, serving the loaf as punishment slashed the number of fights, assaults, and disorderly conduct incidents.

But give people options, and they'll eat a lot. This is commonly known as the buffet or cafeteria effect. Meals with many courses, and foods with a lot of ingredients and a range of sensations, inspire people to keep eating. In 1980, a husband-and-wife team of scientists at Oxford, Edmund and Barbara Rolls, did an experiment in which thirty-two volunteers taste-tested eight different foods: roast beef, chicken, walnuts, chocolate, cookies, raisins, bread, and potatoes. Each volunteer then ate a big helping of one of those foods, and finally, sampled them all again. This time, all of the volunteers gave the food they'd overindulged in a failing grade.

Flavor's deep connections to pleasure drive this phenomenon. The experience of pleasure, at its height during the first few bites, declines as a food is eaten. But eat a different dish, and gratification returns. Rolls and Rolls used electrodes and scans to discern how the brain accomplished this perceptual juggling. As the stomach fills, it sends a series of hormonal "Stop!" signals to areas of the brain that regulate appetite and pleasure. Neurons in these places stop firing. But one location, the orbitofrontal cortex, has neurons that respond to *specific* tastes, smells, and other sensations. It can shut down the pleasure response to one kind of food while keeping it alive for others.

Which is why there's always room for dessert. Main courses usually lack sweets, so even on a full stomach, dessert tastes good. And since the world sugar supply began to relentlessly expand four hundred years ago, elevating dessert to a cooking specialty placed at the end of a meal, habit has ensured hundreds of millions of brains are conditioned to expect and anticipate it.

Potato chip manufacturers had no knowledge of biology when they began to flavor their products with spices in the 1950s. But gradually, the food industry started to develop its own insights into flavor and appetite. Howard Moskowitz was a young scientist working at the US Army's Natick Laboratories, about twenty miles west of Boston, when he confronted the military version of the prison loaf in the early 1970s: the "Meal, Combat, Individual" ration. Used to feed troops fighting in Vietnam, the MCI was a modern version of Charles Darwin's meat tin. It consisted of a portable packet of four canned foods—a meat, such as beef or turkey loaf; bread, crackers, and cookies; cheese spread; a dessert; and an accessory pack with salt, pepper, sugar, gum, and cigarettes. Soldiers hated them. The ham and lima beans entrée was so despised it was considered bad luck to call it by name; to some it was known as "ham and motherfuckers."

Soldiers had become so spoiled by the sensory bombardment of fast food that they disliked even passable military fare—even in foxholes, where there was no alternative. The Natick lab was tasked with fixing this problem. At home, "soldiers were given free food in the commissary or mess, yet they were spending their own money on McDonald's," Moskowitz said. "The Army worried they were spending money on food that might not be nutritious. How do you make military food attractive to the soldiers?"

As a graduate student in the Harvard department of psychology, Moskowitz had worked in the psychophysics lab, experimenting with the perceptions of sweet and salty solutions. (At one point in the late 1960s, Edwin G. Boring, the progenitor of the mistaken tongue map, then recently retired from teaching, took him to dinner at the Faculty Club and regaled him with psychology stories from decades past.) Moskowitz wanted to study the biology of pleasure, but when he proposed it as a thesis topic, his supervisor told him it would never be taken seriously. A scientist's job was to discover universal, unvarying principles in nature, the adviser declared;

the enjoyment people received from flavor was quirky and variable, a scientific nuisance factor better averaged out of the picture.

At Natick, pleasure was the objective. Moskowitz started to look for ways to manipulate enjoyment. He ran taste tests focusing on sugar. He found a consistent pattern: as concentrations rose from zero, pleasure grew, then leveled off, and finally dropped. This wasn't a new or even a surprising discovery: tiny concentrations of sugar, or anything, are barely detectable, and large ones become overwhelming. But the idea that there was a "just right" point in the middle had never been applied to mixtures of many ingredients, or to foods such as ketchup. A colleague at the lab eyed the peak of one of Moskowitz's pleasure graphs and remarked, "Howard, you have discovered the bliss point."

Moskowitz began combining ingredients in different proportions and running more taste tests to find the bliss point for each mixture. Some had more than one: pleasure rose, then fell, then rose again as potency increased. Not everybody had the same bliss points, he noticed. There was a kaleidoscopic range of perception and sensitivity. This work helped to shape the improved flavors of the next generation of the military ration, the "Meal, Ready-to-Eat," or MRE, its courses now sealed in lightweight plastic pouches. But Moskowitz thought he had discovered something that had applications far beyond rations. He pitched his ideas to some food companies. At first, none was interested.

The problem was another old prejudice. "Fundamentally, they thought everybody was pretty much the same," Moskowitz said. "They realized there were differences from person to person; it was known in the literature since the 1890s. But they said, 'Those differences, let's sweep them under the rug.

We know people are different, but we don't really know what to do with it.' There were no organizing principles."

Food companies recognized his qualifications and offered him work as a consultant anyway, and he left the Natick lab. He combined research about bliss points with detailed surveys of flavor preferences, seeking to find the tastiest blends of contrasting ingredients. Certain rich foods were intensely delicious, but the taste became overwhelming. "Can you imagine eating hamburger every day? I think so. White bread? I think so," he said. "Steak? I don't think so. Duck? Maybe you like it, but can you imagine eating Peking duck every day of your life?" Blander flavors produced by mixes of contrasting ingredients triggered pleasure and gratification without overdoing it. Ketchup, for example, contains liberal doses of umami, sweet, salty, and sour tastes, with pleasant aromatic notes. Potato chips are rich and vivid, but melt away quickly. To target consumers with different preferences, recipes could be formulated several ways.

In 1986, Moskowitz persuaded Maxwell House, which was losing a war for the coffee marketplace to Folgers, to start selling a selection of roasts: weak, medium, and strong. Sales rebounded. He told Campbell's, whose Prego spaghetti sauce had failed to make inroads against dominant Ragu, to make an extra-chunky version. It earned them $600 million. Moskowitz also found that the powerful hedonic spells of sweetness, saltiness, and fat could raise the bliss points of many foods. Companies began spiking products with them. Moskowitz had helped to lay the groundwork for the dazzling, sometimes bewildering, landscape of taste consumers now navigate.

• • •

Flavors bloom from the bottom up: from the fizz of chemical reactions on the tongue and through metabolic signals from the gut. But our over-developed frontal cortices, which handle thinking and decision making, mold them from the top down, too.

The color, appearance, and identity of a food is as important as any ingredient. Researchers at the Nestlé Research Center in Lausanne, Switzerland, tested this idea. They showed volunteers lying in an fMRI scanner pictures of high-calorie dishes such as pizza and lamb chops, and low-calorie foods such as green beans and watermelon. Volunteers' tongues were stimulated by a small electrode, producing a mild, neutral taste. As they viewed the pictures, their perceptions of this taste changed. The healthy foods got only mild responses. But the brain lit up at the sight of the rich foods, with activity focused around the orbitofrontal cortex. The electric taste instantly became yummy. The results showed that a pleasant taste takes shape as one lays eyes on delicious food: the mere suggestion of pizza tastes good. It's a complicated response that weaves together vision, memory, and knowledge. It's also nearly instantaneous. "These reactions happen within hundredths of a second: these initial representations are activated, and then there is a change from sensation to cognition," said Johannes le Coutre, the neuroscientist who directs Nestlé's lab. "The visual and gustatory signals integrate with each other."

Many other cognitive impressions influence flavor, including the weight, shape, and color of bowls and utensils. A light plastic spoon makes yogurt seem denser and more expensive than a heavier one. A blue spoon makes pink yogurt taste saltier. A white spoon improves white yogurt, but makes pink yogurt taste worse. Salty popcorn tastes sweeter in a blue or red bowl, while sweet popcorn tastes saltier from a blue bowl. The subtle flavors of wine are especially vulnerable to such manipulation. When people are told a wine is more expensive, it tastes better. The color of a wine, indelibly linked to certain flavors, generates the biggest biases. In a 2000 experiment, students from the oenology program at the University of Bordeaux taste-tested a white wine made from semillon and sauvignon blanc grapes, a red mix of cabernet sauvignon and merlot, and finally the white wine again, this time dyed to appear red. They then wrote down their impressions. Their descriptions of the dyed white wine included many flavor descriptors usually used for reds, including "chicory," "coal," and "musk."

Food companies exploit such tricks to nudge consumers' senses and buying habits. Names, colors, and packaging are designed to leap out at shoppers from supermarket shelves, preternaturally forming a taste in the mind. The most powerful of these tools is, by far, the brand. It's an entire set of memories, feelings, and associations organized around a single name and a logo, which the brain accesses each time it's glimpsed.

In 2004, scientists at Baylor University in Houston ran experiments to show exactly how brands make such deep impressions. They studied reactions to Coke and Pepsi. The two colas have similar chemical makeups and flavors, colors, and consistency. In a taste test with no labels, the volunteers rated both about equally delicious. But once the drinks were labeled, Coke won hands down. Labeled Coke also beat unlabeled Coke. The Pepsi brand, meanwhile, did not seem to influence taste: labeled Pepsi scored the same as unlabeled. Next, the volunteers were placed in an fMRI machine, where they prepared to sip sodas pumped through three-foot-long straws. Before the sip of cola arrived, a screen flashed an image of a Coke or a Pepsi can. Pepsi's performance was again underwhelming. When the Coke can appeared, the hippocampus, the memory hub, came to life before the volunteers sipped. So did an area of the prefrontal cortex tied to conscious perception. The mind appeared to be accessing the vast trove of Coke's cultural associations and setting expectations, bringing experience to bear on flavor.

As the twenty-first century arrived, food and beverage manufacturers had reached the limits of their ability to manipulate the senses. Every possible trick had been tried. Choices proliferated. Swinging through ancient jungles, our monkey ancestors grabbed at luscious ripe fruits amid the greenery. But what happens when the fruits outnumber the leaves?

Draeger's, an upscale supermarket in Menlo Park, California, is known for its dizzying specialty food selections: 250 kinds of mustard, 75 kinds of olive oil, and over 300 different jams. In a now-famous experiment done in 1995, Sheena Iyengar, a professor of business at Columbia University, and Mark Lepper, a psychology professor at Stanford, tested how much choice was too much. In alternating hours, testers dressed as Draeger's employees invited customers to taste from one of two samplers: one with twenty-four kinds of jam, the other with six. After tasting, the customer was given a one-dollar-off coupon to buy a jar. Customers tasted an average of one or two jams at both displays, though more visited the large selection. But in the metric that mattered—sales—the smaller display won hands down. Nearly a third of the people who sampled from it bought jam. Only 3 percent of those who tasted the twenty-four-jam cornucopia bought any afterward. The volume of choices was a deterrent.

Choosing a jam to buy is more complicated than snatching a fruit from a tree. Flavor plays a central role, but so do brand, cost, and whether it will be spread on bread or English muffins. The human frontal cortex, a synapse away from areas that process the senses, memory, pleasure, emotion, and voluntary movement, was built for complex decision making. It pulls those threads together, assessing costs and benefits, gaming future scenarios, nudging the whole brain to action. In fMRI scans of people weighing a choice of foods to buy, scientists have found a small area of the right medial orbitofrontal cortex consistently working hard during the moment of decision. Every new flavor formulation, and every multimillion-dollar ad campaign, is ultimately about getting certain neurons in this spot to fire. But add more and more choices and the decision grows more difficult: there's simply too much information to process. At a certain point, no new taste, no matter how seductive, can cut through the clutter.

Of course, these obstacles didn't slow the march of taste innovation, which relied more and more on new technologies that would make Howard Moskowitz's methods seem as quaint as Joe Murphy's kitchen-table potato chip spice formulations. In the mid-2000s, Kyle Palmer, a biologist who cofounded a taste research startup, Opertech Bio, devised a new method of flavor testing: he replaced humans with rats. Humans have delicate palates and formidable powers of expression and rats are garbage-eating vermin, but the similarities are greater than most people would like to admit. Rats have lived off the waste of human societies for thousands of years. And though they cannot describe flavors, they do have measurable reactions to food that Palmer found a new way to exploit.

He put his rats in Skinner boxes designed to let them sample dozens of liquid flavor formulations out of small, dimpled trays. Pleasure was measured by the number of times they licked from each mixture. They'd typically lick a sugar solution thirty times; water, about twenty; a bitter mix, only once or twice. With this baseline, any flavor compound could be rated by the number of licks. The rats were also trained to evaluate. If they were testing would-be sweeteners, they pressed one lever if a sample tasted like sugar, a second if it did not. After each sampling, they got a—tasteless—food pellet as a reward. Opertech rats spend their whole lives this way, living out a normal span of about three years; their minders get to know their personalities, and their tastes.

This may seem like a crude way to evaluate flavors, but mindlessness is its greatest virtue. To develop new ingredients, food companies sift through endless lists of chemicals, hoping to find even one with potential. By combining tasting with machine-like repetition, Palmer's system generated impressive amounts of data in a short time, substituting brute force for nuanced observation. Four trained rats could test thousands of candidate compounds over a few days, pinpointing the tastiest for further study.

A competing biotech firm, Senomyx, took this approach a step further, using human tissue in place of lab rats. Senomyx scientists decoded, and patented, the DNA for sweet, umami, and some bitter human taste receptors. (Yes, taste genes and other parts of the human genome can be patented.) They infused these DNA strands into a line of kidney cells used in cancer research. The DNA went to work making taste receptors, giving Senomyx an unlimited supply of taste cells in petri dishes. These can be dosed with new flavors, the sub-tleties and flaws of each gauged down to the molecular level. "We can identify when we have a flavor agent that has a bit-terness component to it, determine what bitter receptor it's acting on, and basically dial out the bitter off-taste," said David Linemeyer, a vice president at Senomyx. This system makes Opertech's rats look lazy; it can do in hours what takes them days.

The Senomyx approach had one drawback. The host cells used were descended from stem cells taken from a human fetus that was aborted in the early 1970s. Since then, these cells have been a fixture in medical and biotech research. But Senomyx's job was to develop flavors. However remote, an association with abortion would be devastating to a food or drink brand. Anti-abortion groups found out about this in 2011 and began to protest; a bill was introduced in the Okla-homa state legislature to ban any food developed using this technology. Senomyx, then working with Pepsi, promised not to use that cell line in its soft-drink research.

Both Senomyx and Opertech trained their technology on the hardest taste problem of them all, finding a truly sugary sugar substitute, and both hit on a similar potential solu-tion. Opertech's rats liked a compound called Rebaudioside C, or Reb C for short, a derivative of the stevia leaf. (The ste-via extract already used in many products is a related com-pound called Reb A.) Reb C itself wasn't sweet; however, it made sugar taste sweeter. With such a sweetness enhancer, soft drink makers could reduce the amount of sugar in a drink while maintaining the authentic taste. In 2013, Senomyx and Pepsi announced they had found a similar compound. However revolutionary, though, it was not clear the public would embrace this approach: "Americans Will Be Drugged to Believe Their Soda Is Sweeter," read a headline on the website *Gawker.*

Pepsi researchers devised another way to enliven the mundane experience of sipping a soda. It was based on a straightforward premise: people form judgments about food by sniffing it first. Experience teaches people that delicious smells precede delicious food: the warm, bracing aroma of coffee, the smell of bacon crackling in a frying pan, the scent of chocolate chip cookies fresh out of the oven. Pepsi's "aroma delivery system," patented in 2013, was a gelatin capsule, less than half a millimeter across, containing an aroma designed to create a cola or citrus imprint on the senses and brain before the first swig. Twisting open the cap would break the capsules, releasing the pleasing smell. A signature fragrance might evolve into something like the sound of a pop-top can, the signal of a thirst-quenching drink. This is not the only recently discovered way to exploit the associative powers of aromas; in 2013 a Japanese company began selling a smartphone accessory and app that released pleasing smells into the air, including coffee, curry, strawberry, and Korean barbecue.

Food technologies have begun to dispense with nature itself, and with it, food and flavor traditions that reach back thousands of years or more. At an event in London in 2013, Dutch scientists staged a tasting of the world's first hamburgers made from meat grown in a lab. The ecological costs of raising cattle are high; making meat without them might one day free up land for other purposes, reduce the beef industry's environmental impacts, and feed millions. The research was funded by Google cofounder Sergey Brin, who provided a grant of $330,000; the project's creators hoped to scale up and grow meat for the market within a decade or two.

The scientists, led by Mark Post, took adult stem cells

from cattle, then grew them in a culture of antibiotics to prevent microbes in the surroundings from infecting them. They used serums derived from calf and horse fetuses to spur growth and make the stem cells develop into the right kind of muscle tissue. After a few weeks, the small clumps of cells were put into petri dishes. They grew into fibers that knotted together to form small strips of muscle, each about a centimeter long. To add bulk to the muscle tissue, the scientists stretched the meat over a scaffold made of soluble sugar. The stringy strips were then compacted into pellets and packed en masse to make a burger. The finished product comprised 20,000 strands of meat, each containing 40 billion cow muscle cells, along with bread crumbs and a binder added to hold them together. The burgers were cooked in a pan with sunflower oil and butter.

Real beef is red, fatty, and delicious. It grows marinating in a mixture of blood, natural hormones, and amino acids, and carries the imprint of an animal's diet and experience. The lab burgers were white, and had to be colored with a mix of saffron and beet juice. They didn't taste like meat—or like much of anything. Hanni Rützler, a food scientist, described it in pointedly non-meaty terms: "crunchy and hot" and "a bit like cake." Post planned to add lab-grown fat (which can also be grown from stem cells) to later versions. The burgers may one day approach edibility, or taste good. But they are unlikely to rival the taste of real meat anytime soon.

Sometimes, food and flavor part ways completely. In the early 2010s, Rob Rhinehart, a Silicon Valley software engineer, became fed up with eating. Whatever pleasure he received from savoring tasty foods, or just filling his stomach, was outweighed by the constant inconvenience. He resented having to shop, cook, and wash dishes. He didn't want to go to a restaurant or wait for takeout to arrive. Rhinehart was also suspicious of most of the food he ate. He knew only that it tasted good, and in truth was probably unhealthy.

The straightforward way to rebel against the tyranny of the food system is to return to natural ingredients and fresh, simple flavors. As author Michael Pollan put it: "Eat food. Not too much. Mostly plants." There are many ways to do this, some not especially tasty. They include the Paleolithic diet, based on the idea that our genes and bodies are better suited to eating food that could have been acquired by hunter-gatherers, such as lean, grass-fed meats, milk, eggs, fruits, and nuts.

But Rhinehart was a tech guy, not a foodie or hipster, and he decided to use technology to create the perfect food, building it from first principles. He researched the human body's nutritional needs and collected the most basic chemical ingredients available—the only recognizable elements were olive oil, fish oil, and salt. The contents were carbohydrates, proteins, fats, cholesterol, sodium, potassium, chloride, fiber, calcium, and iron, and a long list of vitamins and other nutrients. When blended together, they looked something like a milkshake, with a faint coffee-ish tint. To some, it resembled vomit. "At the time I didn't know if it was going to kill me or give me superpowers," Rhinehart wrote. "I held my nose and tentatively lifted it to my mouth, expecting an awful taste. It was delicious! I felt like I'd just had the best breakfast of my life. It tasted like a sweet, succulent, hearty meal in a glass."

He called it Soylent, after the 1973 movie *Soylent Green*, set in a dystopian future New York City where the only food available is the eponymous green wafer, supposedly made from processed plankton. The lie is exposed in the movie's final words: "Soylent Green is people!"

Rhinehart ingested nothing but Soylent and water for a month, making himself the guinea pig in an ongoing scientific experiment. He monitored his weight and drew blood daily to test for several important nutritional markers, fine-tuning the Soylent recipe as he went along to ensure he was getting the right balance of nutrients. At one point, his potassium level rose and his heart rate with it; he felt faint, so he reduced the concentration of potassium in the shake. When he began to lose weight, he drank more of it. The formulation cost $154.82 a month, plus shipping for the ingredients; previously, grocery shopping and eating out had cost him close to $500 per month. Pitching this combination of economy, nutrition, and time-saving helped him raise $1.5 million in a Kickstarter fund-raising appeal. He won another $1.5-million infusion of venture capital from Silicon Valley entrepreneurs to help bring Soylent to market.

Chemically, Soylent was not much different from the nutrient concoctions used in feeding tubes for decades. Nor did Rhinehart's self-experimentation show much about how other people, let alone masses of consumers, would respond to a Soylent diet. But its effects on his food and flavor experiences were revealing. He felt sharper. He was never hungry, and no longer craved the junk food he had occasionally binged on before. Meeting the body's nutritional needs so precisely, Rhinehart thought, had addressed the central problem of the junk food age, offering relief from the body's constantly over-revved cycles of craving and gratification—a kind of biological reset. But then the monotony problem struck; bland shakes became a chore to drink. After his initial experiment, Rhinehart continued drinking them, but also allowed himself to indulge in old-fashioned eating and drinking a couple of times a week. He spiked his shakes with vodka. He ate sushi regularly, and came to appreciate its delicate flavors and the craft of the sushi chef, which he now took the time to observe. Only by giving up food was he able to appreciate it.

The apotheosis of this food trend, perhaps decades away or longer, is virtual flavor. When Nimesha Ranasinghe, a computer scientist in Singapore, studied virtual realities, he noticed something missing. Sophisticated head- and hand-sets could trick the eyes, ears, and even the skin into a feeling of immersion in a fabricated digital space: a spaceship, an alien world, ancient Rome. But without flavors, virtual reality would always be an incomplete and impoverished experience. Ranasinghe took a variant of the tongue electrodes used by Nestlé's scientists and experimented with them to see if he could create tastes out of nothing but a mild electric current. He made a device he named the "digital lollipop": a small sphere containing one electrode rests on the tongue; a second electrode is in contact with the tongue's underside.

By subtly adjusting the current's magnitude and frequency, along with temperature, he was able to induce sweet, salty, sour, and bitter sensations directly on the tongue (though not umami). These were crude, but Ranasinghe hoped to refine them, and to develop the means to simulate aromas in hopes of one day creating fully realized virtual flavors. He made digital records of the "tastes," turning them into sequences of ones and zeroes that could be stored on a computer and transmitted over the Internet. Anyone with a digital lollipop device could download the file and "taste" it himself. As the technology improves, a chef may one day be able to create an entire meal, write its flavors to a digital format like a song or a movie, and share it with the world.

Soylent and virtual tastes separate flavor from food. Tasting and savoring become ends in themselves; a form of recreation, play, and art, a feast for the brain and mind without any negative consequences for the body. But can flavor ever truly be liberated from its connection to the body? It draws its power from the gut's metabolic furnace, the call of ancient, implacable drives. Make those redundant, and flavor loses its essence. Taste has wreaked a lot of havoc in the industrial age. But every attempt to undo the damage—to water tastes down or to otherwise trick the senses—has produced unsatisfying results. This is the great catch-22 of flavor. Its great reward is the suggestion of self-indulgence, the whiff of danger from going too far.

CHAPTER 9

The DNA
of Deliciousness

O ne day in 2010, chef David Chang decided to make
some *katsuobushi*, the cured fish flakes that are a staple of traditional Japanese cuisine. *Katsu* means "bonito," a
type of tuna, and *bushi* "pieces" or "shavings." When placed
in broth made with miso, kelp, and tofu, *katsuobushi* turns
into delicate, rippling cellophane ribbons. Its complex flavors
emerge in a process similar to that used for Icelandic *hákarl*,
though more intricate. Thick cuts of bonito are smoked, then
inoculated with molds and packed in dry rice. Over a period
of months, the molds grow, then desiccate. They are scraped
off, only to grow back again. Mold microbes infuse and dine
on the fish flesh, making a suite of aromatic molecules that
rival those of the subtlest cheeses. Amino acids with a strong
umami signature also flourish, harmonizing these diverse flavors. The final product is a solid block, its surface mottled in
greens and blues, ready to be grated.

Chang had already mastered this temperamental fermentation process, having made *katsuobushi* for his five Momofuku
restaurants in New York City. But now he was in the Momofuku Culinary Lab, a cramped, 250-square-foot working

215

kitchen whose sole purpose was experimenting—playfully—with culinary tradition.

Chang and his two partners, chefs Dan Felder and Daniel Burns, debated how to alter this particular recipe. They could fiddle around the margins—adjusting the heat applied during the drying and aging process, perhaps—or they could do something truly subversive and substitute pork for fish. In traditional Japanese kitchens, where *katsuobushi* has been made for three hundred years, this idea would be a culinary oxymoron. They chose to go with it.

As a practical matter, swapping out fish for pork made sense. Bonito and bluefin tuna, which is also used to make *katsuobushi*, are expensive and overfished. In Japan, bluefin tuna are so prized and in such short supply that some have been auctioned for over $1 million apiece. Pork is cheap and plentiful, and pigs can be raised organically. If the partners' experiment worked, they would have a provocative take on a Japanese standard, save money, and minimize environmental harm.

They steamed, smoked, and dried a pork tenderloin, then packed it in raw sushi rice to age. No molds were added; they let the microorganisms on the meat grow instead. Six months later, the aged and petrified pork looked like a Jackson Pollock painting, layered with greens, whites, and coppers—in regular *bushi* preparation, a sign of success. They dubbed their concoction *butabushi*, substituting the Japanese word for pork. But as they were about to taste the finished product, they realized they had a serious problem.

Using pork had seemed like a straightforward ingredient swap, but the process of making *katsuobushi* is rigorously managed, all variables accounted for, the result of centuries of trial and error. By introducing an unknown element, they had profoundly altered the microbiology of that process. The chefs did not know the identity of the molds growing on their cured pork block. They could be toxic, a threat to public health. Even if benign, they could infect ingredients they came in contact with in the kitchen. Even the best-case scenario—that the *butabushi* was safe and its flavor good—was a chef's worst nightmare: they wouldn't know how to re-create it. Natural microbial communities are like snowflakes: each is unique. A different piece of pork would have its own distinctive population of microbes. Even if the molds were identical, small changes in temperature or humidity could influence them in unknown ways, producing wildly different flavors at the end of the aging process. The chefs could duplicate every variable exactly, but the result might never be the same.

Humans mastered fermentation thousands of years ago, but the scientific understanding of it is still in its infancy. The origins can be traced back to 1856, when thirty-four-year-old Louis Pasteur was dean of the Faculty of Sciences at the University of Lille, in France's northern industrial heartland. The father of one of his students, a local distiller named Bigo, approached him with a problem: the spirits he was making from sugar beets were mysteriously turning sour. He invited Pasteur to come and inspect the vats, drawing him into one of the major scientific debates of the era. Some doubted that the yeast that powered alcoholic fermentation was alive, maintaining the process was purely a chemical reaction. Others believed that yeast consisted of tiny organisms that sprang to life fully formed from rotting food and corpses, a process called "spontaneous generation."

Pasteur literally plunged into the challenge that Bigo had

presented to him. "Louis . . . is now up to his neck in beet juice. He spends all his days at the distillery," his wife and lab assistant, Marie Pasteur, wrote to her father-in-law. He chemically analyzed the sour gunk from the vats. It contained lactic acid, the chemical that gives spoiled milk its unpleasant taste. Through a microscope, he examined samples taken from both good and bad batches. The good ones were swarming with yeasts. In the bad ones, no yeasts were present, but a smaller, rod-shaped organism was multiplying. Pasteur prepared a solution mixing the two. The rod microbes made more acid, which killed off the yeasts. Pasteur had discovered two distinct processes under way in the vats. The first was the intended one: yeasts making alcohol. The second was basically an infection. Bacteria were producing lactic acid—key in techniques for making cheese and yogurt, but inimical to brewing. Fermentation, it seemed, consisted of organisms living, digesting, and reproducing. The two major theories had both been wrong.

Pasteur's adventure was the beginning of modern microbiology, the study of minute organisms ubiquitous in nature. Among them are bacteria, protozoa, algae, and fungi, including yeasts. Pasteur went on to make a series of scientific breakthroughs that exposed the hidden world of microbes and their role in diseases. He established the modern understanding of germs and vaccinations, which have eliminated or controlled many once ubiquitous infectious diseases such as polio and smallpox, saving tens or hundreds of millions of lives over the past century. But despite Pasteur's continuing interest in beverage making (he wrote a book titled *Studies on Fermentation: The Diseases of Beer, Their Causes, and the Means of Preventing Them*), the field has fallen far short in one respect: very little is known about how microbes make flavor. Compared to the threat posed by disease, there has never been much urgency to study the benign biological underpinnings of cheeses or beers.

Which is why places such as the Momofuku Lab, which tease apart traditional techniques to see what makes them work, are so important. That effort is part of a broader reinvention of cuisine that is under way in restaurants and artisanal venues around the world, in which science and technology meld with old-fashioned kitchen intuition to push flavor into new domains.

This trend owes a lot to molecular gastronomy, a culinary movement that recasts cooking as chemistry. Conceived by Hervé This and physicist Nicholas Kurti, molecular gastronomy began in the 1980s, as This collected homespun cooking advice from a variety of sources—eighteenth- and nineteenth-century cookbooks, kitchen lore, old wives' tales—and tested it in a laboratory, often the first time such conventional wisdom had been scrutinized scientifically. One of these truisms, which he called "culinary precisions," was the notion that the skin of a suckling pig crackles more if the head is chopped off immediately after roasting, which This traced to a book by the eighteenth-century French gastronome Alexandre-Balthazar-Laurent Grimod de La Reynière. To test this nostrum—which he thought unlikely to work—This roasted four suckling pigs in a public experiment. To assure consistency, the pigs came from the same litter and had been reared on the same farm. They were cooked over a large outdoor fire for five hours. Two of the four heads were cut off, and members of the audience were invited to do a blind comparison. The skin of the headless pigs was, indeed, crispier. As This studied their carcasses, he realized why: when the pigs came off the fire, moisture evaporating from their flesh saturated and softened the skin. With the head cut off, the vapor escaped and the skin remained crispy.

In the 1990s and 2000s, This gathered scientists and chefs in a series of workshops to discuss the chemistry and physics of cooking, and how these could alter tastes, but also the body, brain, and mind. They began to experiment with both ingredients and the physical processes of baking, braising, frying, and microwaving to create new dishes that stimulated the senses in unexpected ways. This employed liquid nitrogen to make ice cream (rapid cooling produces a uniformly smooth texture) and calculated the perfect temperature to cook an egg (149 degrees Fahrenheit solidifies the white while leaving the yolk soft and smooth). Drawing inspiration from these meetings, haute chefs began to set up their own culinary labs. They juxtaposed unconventional ingredients and emphasized surprise. Spanish chef Ferran Adrià's dishes included mango juice "spherified"—flash-frozen into a sphere—with a kind of salt obtained from algae until it looks like an egg yolk or caviar, and Parmesan cheese spun to the consistency of cotton candy.

"The act of eating engages all the senses as well as the mind," wrote chefs Adrià of Catalonia's elBulli, Heston Blumenthal of The Fat Duck in Bray, England, Thomas Keller of The French Laundry in California's Napa Valley, and the esteemed food science writer Harold McGee. Their 2006 manifesto was an audacious attempt to map a twenty-first-century understanding of deliciousness in eight hundred words. The highest aim of cooking, they wrote, is to bring happiness and contentment. The way to do this is to mesmerize the senses, but in an era of sensory overload and networked information, the usual cooking techniques and traditions, not to mention the secrecy long associated with methods and recipes, no longer work. "Preparing and serving food could therefore be the most complex and comprehensive of the performing arts. To explore the full expressive potential of food and cooking, we collaborate with scientists, from food chemists to psychologists, with artisans and artists (from all walks of the performing arts), architects, designers, industrial engineers. We also believe in the importance of collaboration and generosity among cooks: a readiness to share ideas and information, together with full acknowledgment of those who invent new techniques and dishes."

As early humans learned to manipulate flame and create the first recipes, flavor became the first crucible for culture. Today, deliciousness at its peak is an art form at the bleeding edge of high culture, a gateway to the sublime. The mysterious core of its creation, elaborate chemical reactions and the pulse of microbial life, makes it intrinsically more complex than art, music, writing, or filmmaking. Microbiology, genetic research, and neuroscience are making new tools available to shape sensory experiences, and to challenge and reinvigorate culinary traditions. Such efforts are ambitious, but also necessarily smaller in scale than the technologies overtaking the food industry. They're also influential: since they inherited the mantle of high cuisine from the kitchens of royal courts in the nineteenth century, top-flight restaurants have, directly or indirectly, molded what everyone eats. Julia Child brought elite French cooking to a mass American TV audience; chain restaurants borrow the showy presentations of the elite. If the unusual dynamics of fermentation could be tamed by one test kitchen, others would follow.

The potentially hazardous nature of their moldy pork did not deter the chefs at the Momofuku Lab. "Failure is our bread and butter," said Felder, who was the lab's director from

2012 to 2014. Mistakes cracked open the culinary process, revealing how individual elements functioned, or failed to. To see if they had indeed failed, Felder sent *butabushi* mold samples to Rachel Dutton, a fellow at Harvard's Center for Systems Biology who studied the behavior and genetics of fungi and bacteria. Dutton cultured the *butabushi* molds and extracted their DNA. She ran it through a gene sequencer, then compared the results to a database of known microbial DNA. The sample contained six species of fungi and two of bacteria. To everyone's relief, none of them was dangerous. But the results were odd. She had expected to find *Aspergillus oryzae*, known to play an important role in *katsuobushi*'s flavor profile. *Aspergillus oryzae* is the mold in *koji*, the rice concoction used in Japanese cuisine. Instead, a wild fungus named *Pichia burtonii* predominated. It wasn't something ordinarily found in raw or cured meat. "We don't honestly know where it came from," Felder said. "Whether it was just in the atmosphere or in the kitchen."

The term "terroir" refers to the distinct sense of place that imprints itself on grapes, and ultimately on the flavor of wine: the lay of land and sea, the climate, changing patterns of winds and humidity, the soil chemistry. This includes the influence of the microbiome, the teeming universe of microbes that covers nearly everything in nature, whose composition varies mile to mile, yard by yard, and season to season. Any fermented food has its own terroir, and the *Pichia* fungus would carry a distinctive flavor imprint from the lab's geographical location—lower Manhattan. The partners did not know what this would be like; city-bred microbes could yield terrible tastes. But when they tasted the *butabushi*, its flavor was good: savory, smoky, and funky like the fish version, but distinctly porky, too.

The *Pichia* discovery was a potential watershed. If its flavor-making abilities could be harnessed and exploited, along with those of other distinctly New York microbes, Momofuku could create American forms of Japanese cuisine instead of merely tinkering with the originals. It had taken ancient people centuries to tame microbes; now science might allow them to accomplish the task in months.

The first batch of *butabushi* had been a lark, but Felder, taking over, resolved to proceed methodically. He ran a series of experiments to assess *Pichia*'s flavor-making abilities against the standard, *Aspergillus*. The results were disappointing: like an out-of-shape jogger competing against an experienced marathoner, *Pichia* performed miserably. In one test, Felder inoculated pork and beef with both molds. *Bushi* made with *Aspergillus* was superior in every way—taste, aroma, texture, and consistency—to that made with its rival. *Aspergillus*'s long history as a fermentation agent made it reliable and predictable; it produced consistently nice flavors—surprisingly, even in the unfamiliar territory of a new kind of meat. This was another new way to make *bushi*, and so had to be considered a success. But Felder was disappointed the *Pichia* had failed.

When he tried to re-create the flavor of the original *butabushi*, his efforts brought further grief. It didn't taste the same. "It's not the same environment, it's not the same ecosystem that allowed it to become the dominant catalyst the first time," Felder said. "We had isolated one variable, but not all of the others." In other words, it wasn't the *Pichia* alone that created the flavor the first time but its interactions with the other organisms—the chemical symphony of many metabolisms acting together.

These disappointments still offered valuable insights.

They showed that microbes could not easily be brought to heel, while also suggesting a vast terra incognita existed for flavor and cuisine. "We know so little about endemic micro-organisms," Felder said, "that there's limitless potential for what flavor compounds could be created."

Felder kept up his microbial tinkering. (He also published a scientific paper on this work titled "Defining microbial terroir: The use of native fungi for the study of traditional fermentative processes.") He made chicken *bushi* (a good flavor profile, but poor texture) and, after many failed tries, produced a decent beef *bushi* (a slight iron-liver flavor, but good texture). He swapped ingredients out of traditional Japanese foods to see what would happen: instead of rice, he used spelt, freekeh (a roasted green wheat), farro (a whole form of wheat), rye, barley, and buckwheat. Instead of soybeans, he tried pistachios, cashews, pine nuts, lentils, chickpeas, and red beans. Felder's pistachio miso was green. It took many attempts to get it right, but it has become one of Momofuku's signature foods, the sine qua non of its scientific fermentation efforts. Felder put a dollop of it on a spoon and I tried it. It was sensational: rich, yet light; complex and earthy, yet vivid.

Dutton, meanwhile, expanded her microbial detective work from *bushi* to cheese. Two, sometimes three or more types of fermentation are employed to turn curds, the bland, chunky solids from milk, into a hunk of flavorful cheese. Several distinct communities of fungi and bacteria overlap and interact. And yet, "Cheese is relatively simple," Dutton said. "If you compare the number of species of microbes in the human gut—in the hundreds if not over a thousand—to cheese, there are about ten. But because you have this simplicity and stability, when you make small changes there is huge flavor diversity."

She began to collaborate with Jasper Hill Farm, an artisanal cheese maker in Greensboro, Vermont. Before dawn each morning, the staff pumps the milk from forty-six Ayrshire cows into a three-hundred-gallon vat in an adjacent farmhouse to be made into cheese. The warm milk is immediately seeded with a mixture of lactic acid bacteria, yeasts, and ripening agents. As bacteria begin to break lactose down to lactic acid, the milk turns sour. After about five hours, rennet, the solidifying agent, is added. One morning when I visited, cheese maker Scott Harbour dipped a knifelike tool into the milk, testing for signs the fats were about to condense into curds: a few minutes later the tank contained a shaking, shimmering solid. Harbour and a colleague hand-cut and lifted big hunks with the consistency of aspic onto stainless steel counters. It was mushy and mild, with only the slightest hint of acidity. The cheese makers compressed globs into cylindrical molds, which are set aside and flipped on a schedule so the whey drains out evenly, creating a consistent texture: a three-dimensional canvas for what comes next.

I had brought my twelve-year-old daughter, Hannah, to watch cheese being made. She loved comfort food, and cheese in particular, things with subtler, richer flavors harmonized by umami. If she had her way, her diet would be macaroni and cheese, grilled cheese sandwiches, quesadillas, pizza, and cheese ravioli dosed with Parmesan. It was hard to get her to eat anything else, and her pediatrician became concerned about the lack of variety. Eating less cheese only strengthened its allure, and it became a source of mordant comedy. She adopted "cheese" as a catchphrase and made her online avatar a wedge of Swiss.

At Jasper Hill, they were making a soft cheese named Winnimere. After the whey drains, the semi-solid cheese cylinders, about five inches in diameter, are cut into small wheels. In the basement, the next phase was under way: cheese makers wrapped a narrow strip of spruce bark around each wheel. They handed Hannah a hat and apron, and she started wrapping and snapping the bark in place with rubber bands. The bark helps it keep its form, while imparting a sappy, woody flavor and a set of microorganisms to the surface. As the cheese ages, these microbes—mainly penicillin molds—form a hardening rind with a mushroomy flavor. Sometimes a virus gets into a batch, infecting both molds and bacteria. The rind turns yellow and the flavor acrid, sour, and oniony. The molds also work their way inside the cheese, joining the lactic acid bacteria. Depending on the mix of microbes at each point inside, the flavor varies millimeter by millimeter.

Winnimere develops a signature sheen of pink and orange that is important for its brand identity, and catches the eye from crowded display cases. Dutton was trying to understand what, biologically speaking, those colors do. Her basic technique was straightforward: she grew cheese cultures in petri dishes, combined them, and observed their behavior. When she uncapped a dish, sharp, funky smells emerged with no cheese present. She walked us into a cool room where test cheeses are stored, dollops of curd arranged in a grid of tiny plastic wells, infused with different combinations of molds and bacteria. On one of them, a bright green penicillin mold grew next to a yellow colony of *Arthrobacter*, a common genus of bacteria usually found in soil. Dutton flipped the container over. A patch of bright pink was blooming out from *Arthrobacter* adjacent to another, unidentified mold.

"We want to know what that pigment is," she said. "Why

[*Arthrobacter*] is producing it. Does the pigment do something? Is it producing it to maybe try and harm the mold because it doesn't like that it's growing next to it, or is it just some sort of general protective response?"

Some microbe species develop a mutually beneficial relationship with others to survive. Some just compete. Both kinds of interactions produce distinct colors and flavors. Understanding those relationships could allow cheese makers to fine-tune their microbe wrangling, expanding the range and shadings of taste. But there are many obstacles. Even known microbes interact with a slew of environmental unknowns—as *Pichia* was before it was identified. It might be something on the grass the cows eat, or an airborne germ entering the aging vaults. This adds a dose of randomness to every batch. Jasper Hill had plans to follow the technique Dutton brought to Momofuku, identifying the source of its homegrown molds and bacteria. Most American cheese makers get their cultures from European manufacturers. Knowing the local microbiome would enable them to patent a distinct Vermont terroir.

To produce that terroir, aging must be carefully managed so that the microbes flourish in just the right way. Zoe Brickley, who oversees this process, took us inside a Jasper Hill vault. A low ridge had been excavated into seven caves, each projecting into the earth at a different angle off a central axis. Naturally cool and moist, they allow Jasper Hill's cheese makers to orchestrate the emergence of flavor over weeks and months with environmental fine-tuning. Their temperatures range between 49 and 53 degrees, and the humidity is kept at 98 percent.

Wheels of clothbound cheddar, eighteen inches across and six inches high, were stacked on towering shelves. Virgin wheels are pressed into a burlap band. Then circles of cloth are applied to the top and bottom and rubbed with lard. This keeps them from drying, and also creates a home for molds, which multiply and turn the cheeses fluffy as the months pass (they're vacuumed and scrubbed before they depart). Tiny bugs called dust mites burrow into the lard, exposing the burlap to air, which helps maintain the right balance of moisture on its surface. Ripening takes roughly a year: the cheeses on the shelves ranged from new to thirteen months old.

The air was loamy and thick, with a hint of ammonia and a floral scent from the mites. Flavors grew and morphed incrementally, each wheel on its own lonely trajectory. Brickley pulled one down, took out a small cheese-tapping tool, stuck it into the bottom, and extracted a sample. Unlike a typical mass-produced sharp cheddar, with a bitter tang and a hint of sulfur, this one was sweet, with a brothy, umami note. But it was also crumbly, a bad sign. "It's sandy, it has a broken quality to it," Brickley said. "I don't see how that can get much better." Bacterial fermentation had run wild inside this wheel, making more acid and reducing calcium and other minerals essential to a smooth texture. The next wheel was only a week younger, but it was completely different; like a civilization unto itself, each wheel's microbial community rises and falls in its own way. This cheese was smooth, with a hint of pineapple flavor. Still, something was missing. "There's not enough meat," she said. "I think it'll get more meaty, like white miso. I think of meats in a flavor, with white miso being the lightest, then chicken broth, then maybe pork, then maybe red meat broths."

Jasper Hill was founded by two brothers, Andy and Mateo Kehler. They have backgrounds in sustainable agriculture, but turned their attention to the microscale ecology of deliciousness. At one end, molds and bacteria battle; at the other, the pleasure networks of the brain respond. They know both are equally capricious. "I had a conversation with my six-year-old son last night. He's turned into a really picky eater," Andy Kehler told us. "He says, 'Dad, I'm going to write a book about the best taste.' That was after he ate a couple of capers with mustard. It definitely wasn't his pork chop or his potatoes or the awesome mushrooms that we had last night. It was the caper."

Deliciousness is a slippery concept. It is an ideal, something chefs aspire to create and everyone wants to experience. It can be roughly defined as what happens when food ingredients, preparation techniques, presentation, and the company of fellow diners merge to create a flavor that transcends its individual elements. Deliciousness is not merely tasting good; spices taste good, but are not delicious. As food manufacturers discovered, it requires a degree of complexity and contrast: varying tastes, aromas, and textures that alert the brain's pleasure centers, but also provoke the senses, keeping them a little off balance.

The new wave of culinary science aims to engineer deliciousness. The Jasper Hill staff taste-tests its cheeses and rates them with a detailed checklist. Their trained palates usually reach consensus—it's the near-great cheeses that provoke wild disagreements. To understand why, they turned to data. Jasper Hill began collecting and graphing all its ratings, making databases much like those generated by the white rats in Opertech's Philadelphia lab. The ratings are aggregated into a number called the Deliciousness Factor, or DF. It's a one-to-ten scale, with ten as the best possible incarnation of a particular cheese: a rare achievement. Sevens rank pretty well; sixes and fives are problematic. The data are also broken down to analyze the problem areas. "Spider graphs" show a cheese's ratings for texture, sweetness, saltiness, rind development, and the trajectory of its ripening over time. Each data point spikes outward around an axis. The bigger and rounder the graph, the better the cheese. If something isn't right, the numbers are variable and the graph more jagged. The worst collapse back to the center, like a black hole.

The notion that something sublime like deliciousness can be precisely quantified seems a little absurd. But in the digital age, the world has begun to accumulate vast troves of data that contain hidden patterns of behavior. Graphing cheese flavors as they evolve exposes the flaws in the aging process, or a microbial breakdown, or a period of high humidity. What if recipes—or entire cuisines—could be broken down the same way, their interior dynamics exposed by some digital wizardry?

A good recipe depends on the relationships between ingredients: how they chemically interact and change when mixed together and cooked and how the combination piques the senses. But like the buzzing of microbe metabolisms, recipes are scientific black boxes. Some general principles have been established, such as the Maillard reactions responsible for much of the flavor in cooked food. But the chemical dynamics of individual recipes, developed by trial and error and sips from the stirring spoon, remain obscure. Nicholas Kurti, of the molecular gastronomy movement, once said: "I think it is a sad reflection on our civilization that while we can and do measure the temperature in the atmosphere of Venus, we do not know what goes on inside our soufflés."

Yong-Yeol Ahn, a physicist and computer scientist, had worked on computer models of metabolic processes and the dynamics of the social network Twitter before turning his attention to cuisine. These subjects may seem unrelated, but each is a complex system with millions of moving pieces; their behavior often follows common, underlying principles that can be analyzed and understood.

Like disgust, the definition of deliciousness varies widely from place to place, depending on culture and tradition. Some foods don't travel well: cheese spread west from its birthplace in Turkey, but has never been popular in Asia, and Western palates tend to respond poorly to Asian delicacies such as bird's nest soup. Yet these are more the exception than the rule; travelers can usually find something delicious anywhere, which suggests that many ingredient combinations transcend geography and history. As an astrophysicist might try to study the fundamental forces that govern the structure of space and time, Ahn tried to find the hidden commonalities and differences across the universe of cuisines.

Ahn calculated that there were quadrillions of possible recipes that could be made from existing ingredients in the world. Yet he found only a few million in searching the Internet's largest recipe websites and databases. Vast flavor domains lay unexplored, but perhaps they could be opened up virtually.

Ahn mined cookbooks and recipes from around the world to build a database of 381 basic ingredients and 1,021 flavor compounds. Those are not very large numbers, but what matters in a network isn't the number of individual nodes, but the connections between them. A telephone network with two people has only one connection; with four people, six connections; with ten, forty-five connections. To establish the relationships between ingredients, he looked at their shared chemistry: some were very closely related, others distant cousins. This allowed him to quantify their kinship. He mapped the links across a three-dimensional virtual space.

This grand map of the world's flavor preferences looked

like an array of galaxies. Individual ingredients varied in size depending on their importance. Related ingredients were close together, unrelated ones far apart.

The geography revealed subterranean differences in deliciousness, shaped by history. Foods from Western Europe and North America tended to be monotonous, with many very closely related ingredients such as eggs, butter, and vanilla. Cuisines from East Asia and Southern Europe employed many strongly contrasting, chemically diverse ingredients, including garlic, soy sauce, and rice. Only cuisines of East Asia, Latin America, and Southern Europe overlapped. All three used a lot of garlic, and pairs of them used different combinations of onions, tomatoes, and cayenne pepper. There was no common element between these areas and Western Europe and North America.

A Belgian company named Foodpairing is built on this mapping idea. The company's founder, Bernard Lahousse, says he was inspired by the example of Heston Blumenthal, the chef-proprietor of The Fat Duck, in England. In the 1990s, Blumenthal held meetings with physicists, chemists, and flavorists and asked them for recommendations for his menu. One day, he visited the lab at Firmenich, a flavor company in Geneva. A scientist there noted that some of the chemicals in liver are also emitted in floral scents, particularly jasmine. As the concentration of jasmine aroma rises, the scent takes on a distinctly meaty quality that may help the plant attract insects. When Blumenthal returned to his restaurant, he created a dish of foie gras with jasmine sauce. Later, on a whim, he combined chocolate with caviar. It worked beautifully. When he checked back with the lab, he found both contained high concentrations of amines, partially broken-down proteins that make rich, complex flavors.

Lahousse, a pharmaceutical engineer by training, became interested in the biochemistry of cuisine in the early 2000s. "I would go to chefs, presenting myself, saying 'I'm a scientist, how can you use me?'" he said. He worked with several of them to refine recipes, and was struck by the limitations they labored under. "Ferran Adrià could close elBulli for six months and try thousands of combinations, then pick one. But most chefs are not able to do that." (Not being celebrity superchefs, they must work hard just to keep their restaurants running.)

Lahousse teased apart the contents of fruits, vegetables, chocolates, pita chips, oysters, beef, coffee, vinegar, and wine, among others. He built a database and wrote algorithms that could identify shared aromatic compounds among the foods. The algorithms generate maps similar to Ahn's. Links to potential matches spring off a central hub. Closely related items are placed near one another and their best potential matches. Some of these pairings are to be expected, but others are not. Oysters match up well with kiwi and passion fruit, cucumbers with dark chocolate, and milk chocolate with soy sauce. Foodpairing makes its money doing specialized analyses for restaurants and food companies, but Lahousse has made more than a thousand flavor trees publicly available on its website, so chefs, bartenders, and at-home cooks can peruse them for possible inspiration. He has worked on including tastes, textures, and colors into the mix, adding orders of magnitude of complexity.

After the IBM computer system Watson vanquished human champions on the quiz show *Jeopardy!* in 2011, its inventors applied its cognitive capabilities to other fields. One was cuisine: Watson became the first virtual chef. The system was programmed to mine recipe datasets for promising ingredient combinations, and to tap into a trove of scientific information about their real-life tastes. There are limits to Watson's talent; unlike a chef, it cannot instantly recognize a dish gone wrong and tinker to make it work. To address this problem, the IBM engineers collaborated with chefs at the Institute of Culinary Education in New York, who brought to life recipes such as Swiss-Thai asparagus quiche, the Austrian chocolate burrito, and Belgian bacon pudding. Together, they finessed the dishes and the algorithms behind them; engineering and abstract reasoning paired with experience, intuition, and inspiration to conceive new flavors. Man and machine forged a creative bond.

The connective tissue for many of the best food pairings is umami, which creates synergies between diverse tastes and smells, making thin flavors robust. Umami brings together the bitter and acid tang in cheeses and accounts for the fortifying sensation of chicken soup. This versatility is the reason food companies have embraced it—or rather, its chemical variant monosodium glutamate—as an all-purpose flavoring and salt substitute. But as a taste to be cultivated and crafted, brought fully to bear on the problem of deliciousness, umami remained mostly in the province of Asian cuisine. This couldn't last forever. Umami's flavor may be elusive, but science has exposed its potency; as a basic taste it is biologically similar to sweetness, offering a direct route to the brain's pleasure centers. In the past decade, Western cuisine has received a large umami infusion, and it has started to alter the flavor maps.

Umami wasn't identified as a possible basic taste until 1907. Kikunae Ikeda, then a chemist at Tokyo Imperial University, gradually became convinced there was a mysterious, extra taste orchestrating the flavors in his customary lunch of *kombu dashi*, a brothy soup made from dried kelp and *katsuobushi*. So he purchased twenty-five pounds of kelp and set out to isolate it. Chopping, brewing, and distilling the papery greenish stuff, he was eventually able to precipitate glutamate, a salt of an amino acid, a building block of proteins. Ikeda published a paper on his discovery in a Japanese journal, but it received little attention in the West. It took ninety years before umami receptors were identified. Then, in 2005, Adam Fleischman was eating a hamburger from In-N-Out Burger, the popular Los Angeles–based chain that serves thick slabs of ground beef. The word "umami" popped into his head. It rolled off the tongue, languid and exotic. It didn't seem immediately related to the all-American fare he was eating. But of course, it was: burgers are rich in umami.

Fleischman had heard the word in the culinary circles he traveled as the part-owner of two L.A. wine bars, BottleRock and Vinoteque, and also read about it in cookbooks by Heston Blumenthal and other chefs who were integrating it into their creations. "I was trying to isolate what made burgers, and pizza, so craveable," he said. "You put burgers and pizza and nine other dishes in front of people, and 80 percent of the time they'll go for the burger or pizza." Umami was a common and underappreciated element. Fleischman decided to add still more to concentrate the savory effects.

He visited the Mitsuwa Marketplace, a Japanese supermarket, in Santa Monica, and loaded up on umami-heavy ingredients: soy, miso, fish sauce, *kombu dashi*. He retreated to his kitchen and spent hours mushing them together with ground beef and pork in different combinations, with add-ons such as Parmesan, another umami-intensive food. By late that evening, as he tells it, he had created an "umami burger."

Fleischman cashed out his stake in the wine bars and used the money to start a new restaurant. The timing was good; umami's big moment had arrived. Most people will never hear about the microbial communities growing in the Momofuku Lab, though the umami-heavy flavors they produce might wend their way into the food culture. But Fleischman had positioned himself at the crest of a rising flavor trend as it began to reach a wider public. "The audience for food is so much more sophisticated today than it was ten years ago because they've seen these shows, they know what goes into cooking, and they're curious," he said. "They want to know about how you're doing it." Umami was becoming a concept, a brand. Like "Coke," "umami" taps into the power the brain's cognitive functions have over taste and choice; the word suggests something mysterious, rich, and beguiling. Legally, the basic tastes are generic terms, and the US Patent and Trademark Office is skeptical of attempts to obtain exclusive rights to use them. But Fleischman managed to secure the rights to "umami burger" and "umami café," and has mostly kept the name to himself.

Savoriness and novelty proved a potent combination: over five years, Fleischman opened twenty restaurants across Los Angeles, the San Francisco area, New York, and Miami Beach. He planned to scale up to 150. "We want to expand globally," he said, "not like McDonald's, with one on every corner, but every city will have maybe three." There are Umami Burger seasonings, sauces, and T-shirts, and a spin-off make-your-own-pizza chain.

The standard umami burger has eight known sources of its signature taste: beef, Parmesan, tomato, shiitake mushrooms, caramelized onions, an umami sauce, some umami spice powder, and ketchup. But it's not intended to be an umami sledgehammer. A rich sensation envelops the palate, yet it has none of the thickness of fat; it's gentler, leaving space for other flavors to play. "If something is all umami, it doesn't taste good," Fleischman said. "People aren't eating just straight *kombu* or straight anchovies, but they love anchovies layered in other dishes, like a meat sauce. The science was umami; the art was figuring out how to balance it so it worked together in a package."

But the umami burger also seems to contradict a basic principle of deliciousness and most twenty-first-century food. Umami is about harmony, not contrast; comfort, not provocation. An umami burger doubles, triples, quadruples those qualities. The hamburger, Fleischman said, isn't really the object, as it is in other burger joints. It's a sturdy, familiar vessel that takes diners on an umami trip.

In the quest among chefs, foodies, and food corporations to beguile and seduce, an appreciation for individual flavors is lost. There is too much choice, too much contrast, too much pungency, sweetness, and savoriness. Lior Lev Secarz makes spice blends and delicately flavored cookies in his New York shop, La Boîte. The day-to-day flavor experiences of friends and customers in America distressed him. An old boss of his who visited Japan regularly told him that after he arrived in Tokyo, it usually took three days to shake the daily sensory bombardment of American food and clear his palate enough to savor light, subtle dishes such as sushi. "We drink sodas, we drink liquors, we eat hot spicy food, very acidic food. We drink a lot of coffee," Secarz said. "Our tongues, our palates are destroyed. If I was to serve you a very delicate broth with some lemongrass and a piece of raw bonito in it in America, you'd say, 'Where is the Tabasco or A.1. sauce?' Because you cannot taste anything. But if you were living in Japan, you'd think this was the most flavorful thing that you've ever experienced."

Secarz, who was raised in Israel and trained as a chef in France before turning to spices, is something of an anomaly in a culinary world where machines and molecular reactions are ascendant. He uses a mortar and pestle, bowls, and measuring cups, and his own senses and intuition, to make new flavors from spices that have been in use for thousands of years. "We are getting to the point where it's very hard to distinguish yourself in the cooking industry," he said. "Most things have been done, and what people realize more and more is that you don't necessarily need to invent something new, or have fireworks. If you can serve a good, honest, flavorful food, serving the best ingredients, this is where you can make your statement."

His storefront shop in Manhattan's Hell's Kitchen neighborhood smells like a medieval souk on the Silk Road, full of pepper and coriander. Customers are guided by the aromas—most don't bother to taste-test the spices; a sniff is enough.

The shop is closed in the mornings, and Secarz works alone. He spends part of his time on the Internet, scouring prices and markets for supplies; his blends braid strands of flavor from different parts of the world. "Spices are produce," he said. "They grow somewhere, and somebody is spending a lot of time working so we can have them. There's better years, worse years." Often spices dry up due to natural disasters or upheavals, or for economic reasons. Civil war in Syria had cut off the best supply of cumin, forcing him to look elsewhere. He was getting coriander from Canada because his preferred source, India, kept most of it for its domestic market.

Blending begins with an idea. Sometimes it emerges from a particular need. Sometimes Secarz tries to build on a new ingredient he's found. He writes a list of possible ingredients (blends typically contain nine to twenty-three spices; thirteen is the average). He measures each one, and plans how their aromatic and taste elements will layer over one another, and how their colors will mix. Some will be toasted, roasted, or ground. The flavors from a fine grind can be tasted instantly; a coarse one may need to be chewed, releasing flavors in waves. An initial blend, a rough draft, will then undergo a series of tests: "smelling, touching, cooking, baking, making drinks with it, roasting, searing, grilling, frying, oiling."

A blend called Breeze, flavored with lemon and anise, turns that most mundane of fish, tilapia, into a live wire. His cookies have a similar delicacy. In the Sugar Face biscuit collection, a pale wooden box held an aluminum tin. Daria was a round cookie of orange, curry, and dark chocolate. Desert Rose was made with sesame, salted butter, rosebuds, and cardamom; it tasted like an oasis. Secarz's long experience with spices has taught him some valuable secrets. "If you put a grain of pepper in your mouth and crack it and eat it and drink coffee," he said, "it's as if you put two Splendas in your coffee, or a tablespoon of sugar."

Among the different kinds of culinary establishments, technology and tradition contend most fiercely behind the bar. An evolving cocktail craft has turned many bars into working flavor chemistry labs. Dave Arnold, the proprietor of the lower Manhattan bar Booker and Dax, eschews what has become a post–molecular gastronomy cliché: flash-freezing drinks and fruits with liquid nitrogen. While he does have a nitrogen tank outside, a kind of sentinel for the bar, he sees nitrogen as a tool, not an end in itself. "We want it so the drink comes out just looking like a drink, because here, the point is we're trying to change the way we make things behind the bar, not how you experience the cocktail," he said. "So there's not tiny little frozen balls, and there's not a mound of heaping foam and all this other stuff. Because I don't think that stuff has legs. I don't think the average person wants to change the way they drink. They just want something that's a little unexpected in a format that's very comfortable."

Arnold is a chef, beverage formulator, and polymath. (Part of the Momofuku empire, Booker and Dax is named for his two young sons.) He has no formal culinary training: he has a BA in philosophy from Yale and an MFA from Columbia; but he advanced in the world of food because of his sharp intelligence and eclectic interests. He was the first director of technology at the International Culinary Center in New York, a position created to exploit his fascination with centrifuges, vacuum evaporators, and thermal circulators. He is the creator-founder of the Museum of Food and Drink, a work in progress whose aim is to create a Smithsonian Institution for cuisine. In 2013, he mounted a modest first exhibit consisting of a working vintage puffing gun, a huge contraption used to make puffed cereal during the early part of the twentieth century. He also hosts *Cooking Issues*, an Internet radio show.

One evening, the Booker and Dax menu included a margarita made with mescal, yellow chartreuse, Cointreau, and lime; the typical sweet-salt-sour kick seemed both richer and lighter. A mix of gin and grapefruit juice was made with a centrifuge, which extracts only the clear, light juice from the fruit. A drink called the Sure Bet included a rum blend; *crème de mure*, a liqueur made from black currant berries; toasted almond orgeat, a sugary syrup; lemon; and egg white. But the dominant ingredient was lavender. "If you like lavender, you'll love this," the menu said.

Arnold said he and his bartenders had had a long debate over whether the Sure Bet went too far. It arrived in a daiquiri glass, a milky, pinkish-lavender color with a light froth on the top and an aroma like scented bath soap. It presented itself like a drink-soap hybrid, an unapologetic challenge to expectations. "In our discussion, I was like, this drink is balanced, it's well-crafted, and it tastes good. You don't have to like everything," Arnold said. Customers who might find it "like licking out your mother's bath" should check the list of ingredients before ordering.

The lavender aroma was a bit overwhelming at first, but the flavor was mild compared to the initial whiff. Sipping revealed further complexities, like a curtain being drawn back, the almond balancing the tartness of the lemon. What started out as a possible trigger for disgust became vivid and memorable. Arnold mused on why people found unexpected and strange things delicious. "The highest pinnacles of any culture's gastronomy are based in their weird, fermented and/or bitter, complicated, or conflicted flavors," he said. "Why is that? I don't know, but it's almost universally true. Carbonation is supposed to be a sign that fermentation is going wrong, and yet we're attracted to it. We're told nobody likes rotting things. And yet, if you asked me what I would want to eat right now, it is a cheese that smells worse than anyone's dirty laundry."

Many foods are still engineered for blandness, though— especially fruits and vegetables. The typical supermarket tomato has been bred to stand out in supermarket produce sections. It's a bold red, pleasing to the eye, both plump and firm to the touch. It can be packed up and shipped over long distances and retain its robust profile. But it's just not very tasty. Most of the complex flavors have been bred out of it to service the needs of markets and farmers.

"The biggest problem at the heart of everything is, the growers are paid for how many pounds they pack and put in a box. There is no connection between the grower and consumer in terms of flavor. So the system is set up so there's no incentive for anyone to produce a tomato that tastes good," said Harry Klee, a professor of horticulture at the University of Florida. Klee is trying to unwind the past century of the tomato's history, using scientific methods to re-create the lost, earthier flavors of a simpler time.

Klee scoured farmers markets and the Internet looking for heirloom tomato varieties, which carry flavors of the past. But heirlooms alone aren't enough to create a tasty and popular tomato. It must be easy to grow, move, and sell. Taking into consideration the diversity of types and tastes, Klee aimed to develop a portfolio of a few types that taste good and are inexpensive to cultivate. He and his colleagues collected more than two hundred different heirloom cultivars. Their DNA was extracted and their genomes sequenced. Panels of volunteers convened to sample each one and tie their taste qualities to specific genes. They also compared the heirlooms' flavors to those of mass-produced tomatoes to determine what had been lost and how it might be re-created. "We have a long way to go to make a mass-produced, good-tasting heirloom tomato that costs $1.50 a pound," Klee said. But if he succeeds, perhaps the example will show that the science of flavor is useful for something other than driving a mad, blind rush into the future.

Wines have never faced a problem like that of the mass-market tomato. For centuries a mix of traditions, laws, and regulations has guaranteed their consistency. The appellation system is based on the concept of terroir, in which geography is destiny. In France, a Sauvignon Blanc must come from Bordeaux; champagne must come from Champagne. But the inventors of this system never contemplated climate change, another force that will change flavor in imponderable ways.

Over the past fifty years, the average temperature in French wine country has risen by 4.5 degrees Fahrenheit. This has already altered terroir. Most wines simply taste different than they used to. Heat hastens ripening, and the grapes make more sugar, giving wines a higher alcohol content and bolder flavors. In the short term, that's been good. But one estimate predicts that by 2050 it will be too hot to grow Sauvignon Blanc grapes in Bordeaux. Entire wine industries will shift northward, abandoning the lands that shaped them. New growing areas with very different terroir will open up. In one new wine-growing area in Ontario, Canada, vineyards bury their vines in the winter, and if frost strikes during growing season, they light fires and set up giant fans to blow smoke over the plants. The owners can hardly wait for more global warming.

All cuisine is on the cusp of similar changes from global forces no technological wizardry can hold back. In 2011, a symposium on the future of food at the Nordic Food Lab in Copenhagen featured a sampler of live ants, mayonnaise made with bee larvae, and a fermented fish sauce made instead with grasshoppers and wax moth larvae. Some in the audience didn't bite. But many found the insect-based foods rather tasty. The lab, founded by the Danish superchef René Redzepi, blends culinary art with ecological values. Bug cuisine is a promising frontier. Insects are a largely untapped food resource; they contain a lot of protein, vitamins, and other nutrients. They can be captured or farmed, and their environmental impact is light compared to growing cattle, pigs, or chickens. In a warming world prone to droughts and other ecological catastrophes, eating insects might one day become necessary.

The trick is making them palatable. How to create delicious food out of an ingredient widely considered disgusting? Many societies consume insects, but there's no such tradition in European and American cuisine. Ben Reade, the Nordic Lab's director of research and development, and researcher Josh Evans traveled to Australia, where they sampled ants that store a kind of honey in their abdomen, and to Uganda, where they had a lunch of fried crickets, tomato, onions, and chili peppers. They gathered the finest bug recipes and returned to Copenhagen to experiment and consult with a team of chefs, scientists, and anthropologists recruited to solve the problem.

Flavor now sits at the intersection of all the sciences. It's driven more by forces outside kitchens than in them. But chefs and artisans do have one thing working for them: the mystery at the heart of flavor has never truly been cracked. Science has still not explained how flavor can encompass the whole range of human experience—pleasure, joy, disgust, pain, memory—continually hammering these into something new with each new dish, each sip. The protean quality of the flavor sense will help us adapt to the dietary upheavals accompanying climate change—and to the bioengineered food of the future. But while neuroscientists can map firing neurons and hormonal signals, these efforts amount to crude sketches. The flickering images that fMRI scans associate with flavors, feelings, and moods are just a transient scaffolding for systems of mind and action that scientists have only begun to glimpse.

Acknowledgments

A book project begins as a slight idea, then gathers momentum, support, and assistance along the way as it rumbles to its finish. Thanks to my wife, Trish Clay, and my children, Matthew and Hannah, for providing the inspiration for this book, and for their steady support throughout months of reporting, research, and writing. I'm grateful to my mother, Theresa McQuaid, who passed away midway through the writing process, for a lifetime of love and encouragement, which I continue to rely on. The critical eye of my agent, Kris Dahl, helped to develop a germ of an idea into a full-fledged book project. Constance Jones and Norman Oder provided valuable input in figuring out how to tell the story. Thanks to Colin Harrison for believing in and enthusiastically supporting the book—and for pushing me to get it finished. Liese Mayer somehow turned various messy, clunky drafts into a readable narrative. I tip my hat to others at Scribner who brought the book to light, including Will Staehle and Benjamin Holmes. Many scientists, chefs, and others took time to patiently explain their complex work and views to me, including Dave Arnold, Kent Berridge, Zoe Brickley, Ed Currie, Dennis Drayna, Rachel Dutton, Dan Felder, William Leonard, Kyle Palmer, Jill Pruetz, Danielle Reed, Nick

ACKNOWLEDGMENTS

Ryba, Lior Lev Secarz, and Gordon Shepherd. Ted Janger and Vicki Eastus provided a bed and lively company during several research trips to New York. Thanks to Michael Cahill for getting me into Booker and Dax. Eric Rubin provided cigars and liquor at key junctures. During many late nights of writing, the movie *Team America: World Police* kept reappearing on cable and was a welcome break, so thanks, finally, to Trey Parker and Matt Stone.

Notes

Chapter 1: The Tongue Map

1 *eternal present was truly conscious*: S. S. Stevens, "Edwin Garrigues Boring: 1886–1968: Biographical Memoir," National Academy of Sciences (1973).

1 *a young woman's head*: Harvard University Department of Psychology website, http://www.isites.harvard.edu/icb/icb.do?key word = k3007&panel = icb.pagecontent44003%3Ar%241%3F name%3Dhistoricprofs.html&pageid = icb.page19708&page ContentId = icb.pagecontent44003&view = view.do&viewParam _name = boring.html#a_icb_pagecontent44003.

2 *the differences were very small*: Edwin G. Boring, *Sensation and Perception in the History of Experimental Psychology* (New York: Appleton-Century-Crofts, Inc., 1942), 452.

3 *small differences in perception appear huge*: Linda M. Bartoshuk, "The biological basis of food perception and acceptance," *Food Quality and Preference*, no. 4 (1993): 21–32.

4 *she found very limited variation*: Virginia B. Collings, "Human taste response as a function of locus of stimulation on the tongue and soft palate," *Perception and Psychophysics* 16, no. 1 (1973): 169–74.

4 *all over the tongue*: Jayaram Chandrashekar, Mark A. Hoon, Nicholas J. P. Ryba, and Charles S. Zuker, "The receptors and cells for mammalian taste," *Nature* 444, no. 7117 (2006): 288–94, doi:10.1038/nature05401.

4 *maintains the glass designs work*: Robert Simonson, "House of Glass: How Georg Riedel has changed the way we have a drink," *Imbibe*

NOTES

(January/February 2009), https://imbibemagazine.com/Characters
-Georg-Riedel.

5 *"reach to the heart"*: Plato, *Timaeus,* trans. Benjamin Jowett, MIT Internet Classics Archive, http://classics.mit.edu/Plato/timaeus .html.

6 *more skeptical attitude that persisted*: Carolyn Korsmeyer, *Making Sense of Taste: Food and Philosophy* (Ithaca, NY: Cornell University Press, 1999), 26; Korsmeyer, "Delightful, Delicious, Disgusting," *Journal of Aesthetics and Art Criticism* 60, no. 3 (2009): 217–25; Korsmeyer, "Disputing taste," *TPM* (2nd quarter 2009): 70–76. Korsmeyer's work provides an excellent exploration of this topic.

7 *iron key on a leather thong*: Miguel de Cervantes, *Don Quixote,* trans. Edith Grossman (New York: HarperCollins, Kindle Edition, 2009), Kindle location 11884.

8 *explained the irritation they caused*: *Stanford Encyclopedia of Philosophy*, s.v. Alcmaeon, http://plato.stanford.edu/entries/alcmaeon/; Democritus: Stanley Finger, *Origins of Neuroscience: A History of Explorations into Brain Function* (Oxford, UK: Oxford University Press, 2001), 165.

8 *phlegm (earth and water)*: Birgit Heyn, *Ayurveda: The Indian Art of Natural Medicine and Life Extension* (Rochester, VT: Healing Arts Press, 1990), 91–93.

8 *insipid, aqueous, and nauseous*: Finger, *Origins of Neuroscience*, 166.

10 *later won a Nobel Prize*: Nobel Prize website, http://www.nobelprize .org/nobel_prizes/medicine/laureates/2004/.

11 *a hundred thousand times less sensitive*: Nicholas Ryba, interview.

12 *half of a rodent gene for a sweet receptor*: Mark A. Hoon, Elliot Adler, Jurgen Lindemeier, James F. Battey, Nicholas J. P. Ryba, and Charles S. Zuker, "Putative mammalian taste receptors: a class of taste-specific GPCRs with distinct topographic selectivity," *Cell* 96 (1999): 541–51.

13 *such as imagination and emotion*: Mbemba Jabbi, Marte Swart, and Christian Keysers, "Empathy for positive and negative emotions in the gustatory cortex," *NeuroImage* 34, no. 4 (2007): 1744–53, doi:10.1016/j.neuroimage.2006.10.032; Mbemba Jabbi, Jojanneke Bastiaansen, Christian Keysers, "A common anterior insula representation of disgust observation, experience, and imagination shows divergent functional connectivity pathways," *PloS One* 3, no. 8 (2008): e2939, doi:10.1371/journal.pone.0002939.

15 *carrot-flavored cereal*: Julie A. Mennella, Coren P. Jagnow, and Gary K. Beauchamp, "Prenatal and postnatal flavor learning by human infants," *Pediatrics* 107, no. 6 (2001): e88, doi:10.1542 /peds.107.6.e88.

15 *probing, adventurous periods*: Alison Gopnik, Andrew N. Meltzoff, and Patricia K. Kuhl, *The Scientist in the Crib: What Early Learning Tells Us About the Mind* (New York: HarperCollins, 2000), 186.

Chapter 2: The Birth of Flavor in Five Meals

19 *fossil of a predator eating its prey*: Mark A. S. McMenamin, "Origin and Early Evolution of Predators: The Ecotone Model and Early Evidence for Macropredation," in *Predator-Prey Interactions in the Fossil Record*, eds. Patricia H. Kelley, Michal Kowalewski, and Thor A. Hansen (New York: Kluwer Academic/Plenum Publishers, 2003), 379–98.

20 *to kill, and to feed*: University of California Museum of Paleontology website, http://www.ucmp.berkeley.edu/cambrian/camblife.html.

21 *450 million years ago*: Robert M. Dores, "Hagfish, genome duplications, and RFamide neuropeptide evolution," *Endocrinology* 152, no. 11 (2011): 4010–13, doi:10.1210/en.2011-1694.

23 *the brain's basic structure*: John Morgan Allman, *Evolving Brains* (New York: Scientific American Library, 2000), 76. Allman's book provides an excellent exploration of the evolution of the brain over the entire history of life; Helmut Wicht and R. Glenn Northcutt, "Telencephalic connections in the Pacific Hagfish (*Eptatretus stouti*), with special reference to the thalamopallial system," *The Journal of Comparative Neurology* 260 (1998): 245–60; R. Glenn Northcutt, "Understanding vertebrate brain evolution," *Integrative and Comparative Biology* 42, no. 4 (2002): 743–56, doi:10.1093/icb/42.4.743.

24 *250 million years earlier*: Seth D. Burgessa, Samuel Bowringa, and Shu-zhong Shen, "High-precision timeline for Earth's most severe extinction," *Proceedings of the National Academy of Sciences* 111, no. 9 (2014): 3316–21, doi: 10.1073/pnas.1317692111.

26 *three-dimensional images of meteorites*: High-Resolution X-ray Computed Tomography Facility at the University of Texas at Austin website, http://www.ctlab.geo.utexas.edu/.

27 *more quickly and gracefully than its predecessors*: Timothy B. Rowe, Thomas E. Macrini, and Zhe-Xi Luo, "Fossil evidence on origin of the mammalian brain," *Science* 332, no. 6032 (2011): 955–57, doi:10.1126/science.1203117.

29 *occurred in a species of monkey*: Yoav Gilad, Victor Wiebe, Molly Przeworski, Doron Lancet, Svante Pääbo, "Loss of olfactory receptor genes coincides with the acquisition of full trichromatic vision in primates," *PLoS Biology* 2, no. 1 (2004): E5, doi:10.1371/journal .pbio.0020005.

29 *the fruit-vision hypothesis*: B. C. Regan, C. Julliot, B. Simmen, F. Viénot, P. Charles-Dominique, and J. D. Mollon, "Frugivory and colour vision in *Alouatta seniculus*, a trichromatic platyrrhine monkey," *Vision Research* 38 (1998): 3321–27; B. C. Regan, C. Julliot, B. Simmen, F. Viénot, P. Charles-Dominique, and J. D. Mollon, "Fruits, foliage and the evolution of primate colour vision," *Philosophical Transactions of the Royal Society B: Biological Sciences* 356, no. 1407 (2001): 229–83, doi:10.1098 /rstb.2000.0773.

30 *in some primates*: N. J. Dominy, J. C. Svenning, and W. H. Li, "Historical contingency in the evolution of primate color vision," *Journal of Human Evolution* 44, no. 1 (2003): 25–45, doi:10.1016 /S0047-2484(02)00167-7.

31 *most other birds, respectively*: Allman, *Evolving Brains*, 176.

31 *eyes pointing in all directions*: Ibid., 128.

31 *nerve centers for making faces*: Seth D. Dobson and Chet C. Sherwood, "Correlated evolution of brain regions involved in producing and processing facial expressions in anthropoid primates," *Biology Letters* 7, no. 1 (2011): 86–88, doi:10.1098/rsbl.2010.0427.

32 *result of a random wildfire*: Naama Goren-Inbar, Nira Alperson, Mordechai E. Kislev, Orit Simchoni, Yoel Melamed, Adi Ben-Nun, and Ella Werker, "Evidence of hominin control of fire at Gesher Benot Ya'aqov, Israel," *Science* 304, no. 5671 (2004): 725–27, doi:10.1126/science.1095443.

33 *deer, elephants, and other animals*: Nira Alperson-Afil, Gonen Sharon, Mordechai Kislev, Yoel Melamed, Irit Zohar, Shosh Ashkenazi, Rivka Rabinovich, Rebecca Biton, Ella Werker, Gideon Hartman, Craig Feibel, and Naama Goren-Inbar, "Spatial organization of hominin activities at Gesher Benot Ya'aqov, Israel," *Science* 326 (2009): 1677–79, doi:10.1126/science.1180695.

35 *"Dad, I found a fossil!"*: Celia W. Dugger and John Noble Wilford, "New Hominid Species Discovered in South Africa," *New York Times*, April 8, 2010, http://www.nytimes.com/2010/04/09/science/09fossil.html.

36 *nearly perfectly preserved teeth*: Amanda G. Henry, Peter S. Ungar, Benjamin H. Passey, Matt Sponheimer, Lloyd Rossouw, Marion Bamford, Paul Sandberg, Darryl J. de Ruiter, and Lee Berger, "The diet of *Australopithecus sediba*," *Nature* 487 (2012): 90–93, doi:10.1038/nature11185.

38 *weaker, finer muscles*: Hansell H. Stedman, Benjamin W. Kozyak, Anthony Nelson, Danielle M. Thesier, Leonard T. Su, David W. Low, Charles R. Bridges, Joseph B. Shrager, Nancy Minugh-Purvis, and Marilyn A. Mitchell, "Myosin gene mutation correlates with anatomical changes in the human lineage," *Nature* 428, no. 6981 (2004): 415–18, doi:10.1038/nature02358.

38 *only a tenth*: William R. Leonard, J. Josh Snodgrass, and Marcia L. Robertson, "Effects of brain evolution on human nutrition and metabolism," *Annual Review of Nutrition* 27 (April 2007): 311–27, doi:10.1146/annurev.nutr.27.061406.093659.

38 *used for chopping and scraping*: Peter S. Ungar, Frederick E. Grine, and Mark F. Teaford, "Diet in early *Homo*: A review of the evidence and a new model of adaptive versatility," *Annual Review of Anthropology* 35, no. 1 (2006): 209–28, doi:10.1146/annurev.anthro.35.081705.123153.

39 *observations of savanna chimps*: Jill Pruetz, "Brief communication: Reaction to fire by savanna chimpanzees (*Pan troglodytes verus*) at Fongoli, Senegal; Conceptualization of 'fire behavior' and the case for a chimpanzee model," *American Journal of Physical Anthropology* 141, no. 4 (2010): 646–50, doi: 10.1002/ajpa.21245. Pruetz, interview.

40 *to cook hamburgers*: E. Sue Savage-Rumbaugh and Roger Lewin, *Kanzi: The Ape at the Brink of the Human Mind* (New York: John Wiley, 1994), 142.

41 *none to do anything else*: Karina Fonseca-Azevedo and Suzana Herculano-Houzel, "Metabolic constraint imposes tradeoff between body size and number of brain neurons in human evolution," *Proceedings of the National Academy of Sciences* 109, no. 45 (2012): 18571–76, doi:10.1073/pnas.1206390109.

41 *large burst of growth*: Richard Wrangham, *Catching Fire: How Cook-*

ing Made Us Human (New York: Basic Books, Kindle Edition, 2009), Kindle location 888.

42 *flavor came alive*: Daniel E. Lieberman, *The Evolution of the Human Head* (Cambridge, MA: The Belknap Press of Harvard University Press, 2011), 399–409. For an excellent discussion of this topic, see also Gordon M. Shepherd, *Neurogastronomy: How the Brain Creates Flavor and Why It Matters* (New York: Columbia University Press, 2012), chapter 26.

44 *the size of the neocortex did*: Allman, *Evolving Brains*, 173; R. I. M. Dunbar and Suzanne Shultz, "Evolution in the social brain," *Science* 317 (2007): 1344–47, doi:10.1126/science.1145463.

44 *environments are always changing*: Richard Potts, interview; Richard Potts, "Evolution and environmental change in early human prehistory," *Annual Review of Anthropology* 41 (June 2012): 151–68, doi:10.1146/annurev-anthro-092611-145754; Richard Potts, "Hominin evolution in settings of strong environmental variability," *Quaternary Science Reviews* 73 (2013): 1–13, doi: 10.1016/j.quascirev.2013.04.003; Richard Potts, "Environmental hypotheses of hominin evolution," *Yearbook of Physical Anthropology* 41 (1998): 93–136.

45 *Mount Kilimanjaro, the highest*: M. Royhan Gani and Nahid D. S. Gani, "Tectonic hypotheses of human evolution," *Geotimes* (January 2008), http://www.geotimes.org/jan08/article.html?id=feature_evolution.html.

Chapter 3: The Bitter Gene

48 *thumbs-down to brussels sprouts*: David Lauter, "Bush Says It's Broccoli, and He Says . . . With It," *Los Angeles Times*, March 23, 1990, http://articles.latimes.com/1990-03-23/news/mn-705_1_barbara-bush.

48 *"Make it cauliflower," he said*: "Bush forced to face green nemesis in Mexico," Reuters, February 16, 2001, http://www.iol.co.za/news/world/bush-forced-to-face-green-nemesis-in-mexico-1.61185?ot=inmsa.ArticlePrintPageLayout.ot.

49 *enter their digestive tracts*: Hanah A. Chapman and Adam K. Anderson, "Understanding disgust," *Annals of the New York Academy of Sciences* 1251 (2012): 62–76, doi:10.1111/j.1749-6632.2011.06369.x.

51 *making darker roasts more bitter*: Thomas Hofmann, "Identification of the key bitter compounds in our daily diet is a prerequisite for the understanding of the hTAS2R gene polymorphisms affecting food choice," *Annals of the New York Academy of Sciences* 1170 (July 2009): 116–25, doi:10.1111/j.1749-6632.2009.03914.x.

53 *into his mouth and winced*: Arthur L. Fox, "The relationship between chemical constitution and taste," *Proceedings of the National Academy of Sciences* 18 (1932): 115–20.

54 *Fox told an interviewer*: J. D. Ratcliff, "It's All a Matter of Taste," *The Herald of Health* (May 1963): 16–17, 25.

55 *three would be purple, one white*: Mendel University in Brno website, http://www.mendelu.cz/en/o_univerzite/historie/j_g_mendel.

55 *in a seminal scientific paper*: Fox, "The relationship between chemical constitution and taste," 115.

55 *detected other qualities*: Linda M. Bartoshuk, Katharine Fast, and Derek J. Snyder, "Genetic Differences in Human Oral Perception," in *Genetic Variation in Taste Sensitivity*, eds. John Prescott and Beverly Tepper (New York: Marcel Dekker, 2004), 1.

56 *the men made themselves scarce*: Nathaniel Comfort, "'Polyhybrid heterogeneous bastards': promoting medical genetics in 1930s America," *Journal of the History of Medicine and Allied Sciences* 61, no. 4 (2006): 415–55.

56 *investigators from the University of Toronto*: Norma Ford and Arnold D. Mason, "Taste reactions of the Dionne quintuplets," *The Journal of Heredity* 32, no. 10 (1941): 365–68.

57 *one-way screens*: Dennis Gaffney, "The Story of the Dionne Quintuplets," *Antiques Roadshow*, March 23, 2009, http://www.pbs.org/wgbh/roadshow/fts/wichita_200803A12.html.

59 *"the world of neglected dimensions"*: C. W. W. Ostwald, *An Introduction to Theoretical and Applied Colloid Chemistry: The World of Neglected Dimensions* (New York: John Wiley, 1917).

60 *thousands of different substances*: Francisco López-Muñoz and Cecilio Alamo, "Historical evolution of the neurotransmission concept," *Journal of Neural Transmission* 116 (2009): 515–33.

60 *Taste receptors*: Receptors are special proteins that first emerged at least 1.5 billion years ago, long before organisms had mouths or brains, as an ingenious solution to a basic problem: microbes needed to tell what was going on around them—to detect nutrients or light, and to avoid toxins. Then, when multicellular life

emerged a billion years later, receptors evolved further. On the outside, the body confronted the flux of the world. But on the inside, systems for digestion, respiration, and other bodily functions had to communicate within and among themselves and the brain. Each new task pushed the humble receptor in new directions, molding chemical structures that do thousands of different things.

60 *pore-like opening at its tip*: Chandrashekar, Hoon, Ryba, and Zuker, "The receptors and cells for mammalian taste," 288; Monell Chemical Senses Center website, Monell Taste Primer, http://www.monell.org/news/fact_sheets/monell_taste_primer.

61 *They dubbed it T2R1*: Jayaram Chandrashekar, Ken L. Mueller, Mark A. Hoon, Elliot Adler, Luxin Feng, Wei Guo, Charles S. Zuker, and Nicholas J. P. Ryba, "T2Rs function as bitter taste receptors," *Cell* 100 (2000): 703–11.

61 *Arthur Fox's bitter gene*: Dennis Drayna, Hilary Coon, Un-Kyung Kim, Tami Elsner, Kevin Cromer, Brith Otterud, Lisa Baird, Andy P. Peiffer, and Mark Leppert, "Genetic analysis of a complex trait in the Utah Genetic Reference Project: A major locus for PTC taste ability on chromosome 7q and a secondary locus on chromosome 16p," *Human Genetics* 112 (2003): 567–72.

62 *man's closest relative, the chimpanzee*: Stephen Wooding, "Phenylthiocarbamide: A 75-year adventure in genetics and natural selection," *Genetics* 172 (2006): 2015–23.

62 *produced identical taste experiences*: Stephen Wooding, Bernd Bufe, Christina Grassi, Michael T. Howard, Anne C. Stone, Maribel Vazquez, Diane M. Dunn, Wolfgang Meyerhof, Robert B. Weiss, and Michael J. Bamshad, "Independent evolution of bitter-taste sensitivity in humans and chimpanzees," *Nature* 440 (2006): 930–34, doi:10.1038/nature04655.

63 *turned out to be a taster*: Carles Lalueza-Fox, Elena Gigli, Marco de la Rasilla, Javier Fortea, and Antonio Rosas, "Bitter taste perception in Neanderthals through the analysis of the TAS2R38 gene," *Biology Letters* 5, no. 6 (2009): 809–11, doi:10.1098/rsbl.2009.0532.

63 *one hundred thousand years ago*: Qiaomei Fu, Alissa Mittnik, Philip L. F. Johnson, Kirsten Bos, Martina Lari, Ruth Bollongino, Chengkai Sun, Liane Giemsch, Ralf Schmitz, Joachim Burger, Anna Maria Ronchitelli, Fabio Martini, Renata G. Cremonesi, Jiri

Svoboda, Peter Bauer, David Caramelli, Sergi Castellano, David Reich, Svante Paabo, and Johannes Krause, "A revised timescale for human evolution based on ancient mitochondrial genomes," *Current Biology* 23, no. 7 (2013): 553–59, doi:10.1016/j.cub.2013.02.044; Aylwyn Scally and Richard Durbin, "Revising the human mutation rate: Implications for understanding human evolution," *Nature Reviews: Genetics* 13, no. 10 (2012): 745–53, doi:10.1038/nrg3295.

64 *started making earth ovens*: Richard Wrangham, "Cooking as a biological trait," *Comparative Biochemistry and Physiology—Part A: Molecular & Integrative Physiology* 136, no. 1 (2003): 35–46, doi:10.1016/S1095-6433(03)00020-5.

64 *in a single exodus*: Lev A. Zhivotovsky, Noah A. Rosenberg, and Marcus W. Feldman, "Features of evolution and expansion of modern humans, inferred from genomewide microsatellite markers," *American Journal of Human Genetics* 72 (2003): 1171–86.

64 *crossed at Bab el-Mandeb*: Marta Melé, Asif Javed, Marc Pybus, Pierre Zalloua, Marc Haber, David Comas, Mihai G. Netea, Oleg Balanovsky, Elena Balanovska, Li Jin, Yajun Yang, R. M. Pitchappan, G. Arunkumar, Laxmi Parida, Francesc Calafell, Jaume Bertranpetit, and The Genographic Consortium, "Recombination gives a new insight in the effective population size and the history of the Old World human populations," *Molecular Biology and Evolution* 29 (2011): 25–40.

66 *least bitter-sensitive of early American peoples*: Sun-Wei Guo and Danielle R. Reed, "The genetics of phenylthiocarbamide perception," *Annals of Human Biology* 28, no. 2 (2012): 111–42.

66 *90 percent of non-Africans have it*: Nicole Soranzo, Bernd Bufe, Pardis C. Sabeti, James F. Wilson, Michael E. Weale, Richard Marguerie, Wolfgang Meyerhof, and David B. Goldstein, "Positive selection on a high-sensitivity allele of the human bitter-taste receptor TAS2R16," *Current Biology* 15, no. 14 (2005): 1257–65, doi:10.1016/j.cub.2005.06.042. For a more recent study, see Hui Li, Andrew J. Pakstis, Judith R. Kidd, Kenneth K. Kidd, "Selection on the human bitter taste gene, TAS2R16, in Eurasian populations," *Human Biology* 83, no. 3 (2011): 363–77, doi:10.3378/027.083.0303.

68 *neon colors rather than gentle pastels*: Bartoshuk, "The biological basis of food perception and acceptance," 28–29.

NOTES

70 *bitter receptors might be in the nose*: Robert J. Lee, Guoxiang Xiong, Jennifer M. Kofonow, Bei Chen, Anna Lysenko, Peihua Jiang, Valsamma Abraham, Laurel Doghramji, Nithin D. Adappa, James N. Palmer, David W. Kennedy, Gary K. Beauchamp, Paschalis-Thomas Doulias, Harry Ischiropoulos, James L. Kreindler, Danielle R. Reed, and Noam A. Cohen, "T2R38 taste receptor polymorphisms underlie susceptibility to upper respiratory infection," *The Journal of Clinical Investigations* 122, no. 11 (2012): 4145–59, doi:10.1172/JCI64240DS1.

72 *where they became toxic*: Timothy Johns and Susan L. Keen, "Taste evaluation of potato glycoalkaloids by the Aymara: A case study in human chemical ecology," *Human Ecology* 14, no. 4 (1986): 437–52.

Chapter 4: Flavor Cultures

76 *illegal in the United States until 2007*: Phil Baker, *The Book of Absinthe: A Cultural History* (New York: Grove Press, Kindle Edition, 2007), Kindle location 187–90; Jesse Hicks, "The Devil in a Little Green Bottle: A History of Absinthe," *Chemical Heritage Magazine* (Fall 2010), http://www.chemheritage.org/discover/media/magazine/articles/28-3-devil-in-a-little-green-bottle.aspx?page=1.

76 *trace amounts of thujone*: Dirk W. Lachenmeier, David Nathan-Maister, Theodore A. Breaux, Eva-Maria Sohnius, Kerstin Schoeberl, and Thomas Kuballa, "Chemical composition of vintage preban absinthe with special reference to thujone, fenchone, pinocamphone, methanol, copper, and antimony concentrations," *Journal of Agricultural and Food Chemistry* 56, no. 9 (2008): 3073–81, doi:10.1021/jf703568f.

77 *upper classes a century later*: Harold McGee, *On Food and Cooking: The Science and Lore of the Kitchen* (New York: Scribner, 2004), 759.

80 *kill off other yeasts*: Patrick McGovern, *Uncorking the Past: The Quest for Wine, Beer and Other Alcoholic Beverages* (Berkeley: University of California Press, Kindle Edition, 2009), Kindle location 300.

80 *tens of millions of years old*: P. Veiga-Crespo, M. Poza, M. Prieto-Alcedo, and T. G. Villa, "Ancient genes of *Saccharomyces cerevisiae*," *Microbiology* 150, pt. 7 (2004): 2221–27, doi:10.1099/mic.0.27000-0.

80 *ubiquity of baker's yeast: wasps*: Irene Stefaninia, Leonardo Dapporto, Jean-Luc Legras, Antonio Calabretta, Monica Di Paola,

NOTES

Carlotta De Filippo, Roberto Viola, Paolo Capretti, Mario Polsinelli, Stefano Turillazzi, and Duccio Cavalieri, "Role of social wasps in *Saccharomyces cerevisiae* ecology and evolution," *Proceedings of the National Academy of Sciences* 109, no. 33 (2012): 13398–403, doi:10.1073/pnas.1208362109/-/DCSupplemental. http://www.pnas.org/cgi/doi/10.1073/pnas.1208362109.

81 *tracked these monkey benders*: Dustin Stephens and Robert Dudley, "The Drunken Monkey Hypothesis," *Natural History* (December 2004–January 2005): 40–44.

81 *Dudley suggested*: Robert Dudley, "Ethanol, fruit ripening, and the historical origins of human alcoholism in primate frugivory," *Integrative and Comparative Biology* 44, no. 4 (2004): 315–23, doi:10.1093/icb/44.4.315.

82 *such accidents became recipes*: McGovern, *Uncorking the Past*, Kindle location 449–85. I am indebted to McGovern's fascinating account of primate drinking and the earliest alcoholic beverages.

83 *two, eight, and ten*: Laura Anne Tedesco, "Jiahu (ca. 7000–5700 BC)" in *Heilbrunn Timeline of Art History* (New York: The Metropolitan Museum of Art, 2000), http://www.metmuseum.org/toah/hd/jiah/hd_jiah.htm.

83 *herbs also appeared*: Patrick E. McGovern, Juzhong Zhang, Jigen Tang, Zhiqing Zhang, Gretchen R. Hall, Robert A. Moreau, Alberto Nunez, Eric D. Butrym, Michael P. Richards, Chen-shan Wang, Guangsheng Cheng, Zhijun Zhao, and Changsui Wang, "Fermented beverages of pre- and proto-historic China," *Proceedings of the National Academy of Sciences* 101, no. 51 (2004): 17593–98, doi:10.1073/pnas.0407921102.

86 *ambitious, or maybe just stubborn, project*: Ruth Bollongino, Joachim Burger, Adam Powell, Marjan Mashkour, Jean-Denis Vigne, and Mark G. Thomas, "Modern taurine cattle descended from small number of Near-Eastern founders," *Molecular Biology and Evolution* 29, no. 9 (2012): 2101–4, doi:10.1093/molbev/mss092.

87 *enabling people to digest lactose*: Yuval Itan, Adam Powell, Mark A. Beaumont, Joachim Burger, Mark G. Thomas, "The origins of lactase persistence in Europe," *PLoS Computational Biology* 5, no. 8 (2009): e1000491, doi:10.1371/journal.pcbi.1000491.

88 *8,500 to 7,000 years ago*: Richard P. Evershed, Sebastian Payne, Andrew G. Sherratt, Mark S. Copley, Jennifer Coolidge, Duska Urem-Kotsu, Kostas Kotsakis, Mehmet Özdoğan, Aslý E.

NOTES

Özdoğan, Olivier Nieuwenhuyse, Peter M. M. G. Akkermans, Douglass Bailey, Radian-Romus Andeescu, Stuart Campbell, Shahina Farid, Ian Hodder, Nurcan Yalman, Mihriban Özbaşaran, Erhan Bıçakcı, Yossef Garfinkel, Thomas Levy, and Margie M. Burton, "Earliest date for milk use in the Near East and southeastern Europe linked to cattle herding," *Nature* 455, no. 7212 (2008): 528–31, doi:10.1038/nature07180.

88 *separate curds and whey*: Melanie Salque, Peter I. Bogucki, Joanna Pyzel, Iwona Sobkowiak-Tabaka, Ryszard Grygiel, Marzena Szmyt, and Richard P. Evershed, "Earliest Evidence for Cheese Making in the Sixth Millennium," *Nature* 493 (2013): 522–25, doi:10.1038/nature11698.

88 *the globes begin to clump together*: P. L. H. McSweeney, ed., *Cheese Problems Solved* (Cambridge, UK: Woodhead Publishing Ltd., 2007), 50.

89 *blue-green marbling*: Paul S. Kinstedt, *Cheese and Culture: A History of Cheese and Its Place in Western Civilization* (White River Junction, VT: Chelsea Green, 2012). I drew on this book's thorough account of the evolution of cheese.

89 *closest wild relative*: John G. Gibbons, Leonidas Salichos, Jason C. Slot, David C. Rinker, Kriston L. McGary, Jonas G. King, Maren A. Klich, David L. Tabb, W. Hayes McDonald, and Antonis Rokas, "The evolutionary imprint of domestication on genome variation and function of the filamentous fungus *Aspergillus oryzae*," *Current Biology* 22 (2012): 1403–9, doi:10.1016/j.cub.2012.05.033.

91 *"something really delicious"*: Jean Anthelme Brillat-Savarin, *The Physiology of Taste: Or Meditations on Transcendental Gastronomy*, trans. M. F. K. Fisher (New York: Vintage electronic edition, 2009), Kindle location 1838.

93 *potatoes, meat, and sulfur*: Gerrit Smit, Bart A. Smit, and Wim J. M. Engels, "Flavour formation by lactic acid bacteria and biochemical flavour profiling of cheese products," *FEMS Microbiology Reviews* 29, no. 3 (2005): 591–610, doi:10.1016/j.femsre.2005.04.002.

95 *flavor continually evolves*: Kirsten Shepherd-Barr and Gordon M. Shepherd, "Madeleines and neuromodernism: Reassessing mechanisms of autobiographical memory in Proust," *Auto/Biography Studies* 13 (1998): 39–59.

95 *salty, sweet, sour, bitter, and umami*: Xiaoke Chen, Mariano Gabitto, Yueqing Peng, Nicholas J. P. Ryba, Charles S. Zuker, "A gustotopic map of taste qualities in the mammalian brain," *Science* 333 (2011): 1262–65.

95 *ever-shifting quality of now*: A. D. (Bud) Craig, "How do you feel—now? The anterior insula and human awareness," *Nature Reviews: Neuroscience* 10, no. 1 (2009): 59–70, doi:10.1038/nrn2555.

96 *process pleasure and aversion*: Morten L. Kringelbach, "The human orbitofrontal cortex: Linking reward to hedonic experience," *Nature Reviews: Neuroscience* 6 (2005): 691–702, doi:10.1038/nrn1748.

98 *more powerful sensation*: Clara McCabe and Edmund T. Rolls, "Umami: A delicious flavor formed by convergence of taste and olfactory pathways in the human brain," *European Journal of Neuroscience* 25, no. 6 (2007): 1855–64, doi:10.1111/j.1460-9568.2007.05445.x.

98 *response echoes that for sugar*: Christian H. Lemon, Susan M. Brasser, and David V. Smith, "Alcohol activates a sucrose-responsive gustatory neural pathway," *Journal of Neurophysiology* 92, no. 1 (2004): 536–44, doi:10.1152/jn.00097.2004.

98 *makes tossing back a shot so bracing*: Alex Bachmanov, Monell Chemical Senses Center, interview.

99 *similar to bell peppers*: Amy Coombs, "Scientia Vitis: Decanting the Chemistry of Wine Flavor," *Chemical Heritage Magazine* (Winter 2008–09), http://www.chemheritage.org/discover/media/magazine/articles/26-4-scientia-vitis.aspx.

99 *In a study*: Richard J. Stevenson and Robert A. Boakes, "Sweet and Sour Smells: Learned Synesthesia Between the Senses of Taste and Smell," in *The Handbook of Multisensory Processes*, eds. Gemma A. Calvert, Charles Spence, and Barry E. Stein (Cambridge, MA: MIT Press, 2004), 69–83.

99 *colors associated with words or symbols*: Julia Simner, "Beyond perception: Synaesthesia as a psycholinguistic phenomenon," *Trends in Cognitive Sciences* 11, no. 1 (2007): 23–29, doi:10.1016/j.tics.2006.10.010.

101 *flavor of the food they described*: Jamie Ward and Julia Simner, "Lexical-gustatory synaesthesia: Linguistic and conceptual factors," *Cognition* 89, no. 3 (2003): 237–61, doi:10.1016/S0010-0277(03)00122-7.

102 *myths dating back thousands of years*: Julien D'Huy, "Polyphemus (Aa. Th. 1137): A phylogenetic reconstruction of a prehistoric tale," *Nouvelle Mythologie Comparée* 1 (2013): 1–21.

102 *help themselves*: Homer, *The Odyssey*, trans. Robert Fagles (New

York: Penguin Classics, electronic edition, 2002), Kindle location 5674.

Chapter 5: The Seduction

105 *the sweeter it will taste*: Ayako Koizumi, Asami Tsuchiya, Ken-ichiro Nakajima, Keisuke Ito, Tohru Terada, Akiko Shimizu-Ibuka, Loïc Briand, Tomiko Asakura, Takumi Misaka, and Keiko Abe, "Human sweet taste receptor mediates acid-induced sweetness of miraculin," *Proceedings of the National Academy of Sciences* 108, no. 40 (2011): 16819–24, doi:10.1073/pnas.1016644108.

106 *Radiation damages them, too*: Patty Neighmond, "Chemo Can Make Food Taste Like Metal: Here's Help," *Morning Edition*, NPR, April 7, 2014, http://www.npr.org/2014/04/07/295800503 /chemo-can-make-food-taste-like-metal-heres-help; Marlene K. Wilken and Bernadette A. Satiroff, "Pilot study of 'Miracle Fruit' to improve food palatability for patients receiving chemotherapy," *Clinical Journal of Oncology Nursing* 16, no. 5 (2012): E173–E177, doi:10.1188/12.CJON.E173-E177.

107 *any other people in the world*: Credit Suisse Research Institute, "Sugar Consumption at a Crossroads" (2013), 4.

108 *three-quarters of adults were obese*: Centers for Disease Control and Prevention, "Number (in Millions) of Civilian, Noninstitutionalized Persons with Diagnosed Diabetes, United States, 1980–2011," http://www.cdc.gov/diabetes/statistics/prev/national /figpersons.htm.

110 *exactly what it was*: Asvaghosha, "The Buddhacarita (Life of Buddha)," in *Buddhist Mahāyāna Texts*, trans. E. B. Cowell, F. Max Muller, and J. Takakusu (New York: Dover Publications, 1969), 166; Sanjida O'Connell, *Sugar: The Grass That Changed the World* (London: Virgin Books, 2004), 9.

111 *first dessert cuisine*: Tim Richardson, *Sweets: A History of Candy* (New York: Bloomsbury, 2002), Kindle location 1101–4.

111 *"he becomes invulnerable"*: Richardson, *Sweets*, Kindle location 1125.

112 *the means for refining it*: John Kieschnick, *The Impact of Buddhism on Chinese Material Culture* (Princeton, NJ: Princeton University Press, 2003), 249–54.

112 *sweet water, milk, wine, and honey*: Rachel Laudan, *Cuisine and*

Empire: Cooking in World History (Berkeley: University of California Press, 2013), Kindle location 3085–87.

113 *found in the accounting rolls*: *The Oxford English Dictionary* (Oxford, UK: Oxford University Press, Compact Edition, 1980), 3343–44.

114 *used for food*: Sidney Mintz, *Sweetness and Power: The Place of Sugar in Modern History* (New York: Penguin Books, 1985), 99, 82.

114 *the first cough drops*: *OED*, 2120.

114 *who began planting their own*: J. H. Galloway, *The Sugar Cane Industry: An Historical Geography from its Origins to 1914* (Cambridge, UK: Cambridge University Press, 1989), 63.

115 *horses, cattle, or waterwheels*: Mintz, *Sweetness and Power*, 33–34.

116 *handy for emergency amputations*: Matthew Parker, *The Sugar Barons: Family, Corruption, Empire, and War in the West Indies* (New York: Walker, 2011), Kindle location 1546–48.

116 *obstructed the digestive tract*: Ivan Day, "The Art of Confectionery," in *Pleasures of the Table: Ritual and Display in the European Dining Room 1600–1900: An Exhibition at Fairfax House*, eds. Peter Brown and Ivan Day (New York: New York Civic Trust, 2007).

117 *reading Tryon's writings*: Tristram Stuart, *The Bloodless Revolution: A Cultural History of Vegetarianism from 1600 to Modern Times* (New York: W. W. Norton and Company, 2007), 243–44.

117 *ninety pounds in 1900*: Mintz, *Sweetness and Power*, 67, 143.

118 *"King of Sweets"*: Daniel Carey, "Sugar, colonialism and the critique of slavery: Thomas Tryon in Barbados," *Studies on Voltaire and the Eighteenth Century* 9 (2004): 303–21.

118 *corn syrup became a standard food additive*: Richard O. Marshall and Earl R. Kooi, "Enzymatic conversion of D-glucose to D-fructose," *Science* 125, no. 9 (1957): 648–49; James N. BeMiller, "One hundred years of commercial food carbohydrates in the United States," *Journal of Agricultural and Food Chemistry* 57 (2009): 8125–29, doi:10.1021/jf8039236.

119 *the evolution of complex life*: John H. Koschwanez, Kevin R. Foster, and Andrew W. Murray, "Sucrose utilization in budding yeast as a model for the origin of undifferentiated multicellularity," *PLoS Biology* 9, no. 8 (2011): e1001122, doi:10.1371/journal.pbio.1001122.

120 *"don't you take to drink on that account"*: William James, "To Miss Frances R. Morse. Nanheim, July 10, 1901," in *Letters of William James*, ed. Henry James (Boston: Atlantic Monthly Press, 1920).

NOTES

120 *Olds wrote*: James Olds, "Pleasure Centers in the Brain," *Scientific American* 195, no. 4 (October 1956): 105–17, doi:10.1038/scientificamerican1056-1105.

121 *stimulate itself by pressing a lever*: James Olds and Peter Milner, "Positive reinforcement produced by electrical stimulation of septal area and other regions of rat brain," *Journal of Comparative and Physiological Psychology* 47, no. 6 (1954): 419–27.

122 *flick of a switch*: Morton L. Kringelbach and Kent C. Berridge, "The functional neuroanatomy of pleasure and happiness" *Discovery Medicine* 9, no. 49 (2010): 579–87.

122 *as if licking their lips*: Dallas Treit and Kent C. Berridge, "A comparison of benzodiazepine, serotonin, and dopamine agents in the taste-reactivity paradigm," *Pharmacology Biochemistry and Behavior* 37, no. 21 (1990): 451–56.

123 *"pleasure, euphoria, or 'yumminess'"*: Roy A. Wise, "The dopamine synapse and the notion of 'pleasure centers' in the brain," *Trends in Neurosciences* 3 (1980): 91–95.

124 *alleviate their symptoms*: Alan A. Baumeister, "Tulane electrical brain stimulation program: A historical case study in medical ethics," *Journal of the History of the Neurosciences* 9, no. 3 (2000): 262–78.

124 *the brain of a young man*: Charles E. Moan and Robert G. Heath, "Septal stimulation for the initiation of heterosexual behavior in a homosexual male," *Journal of Behavioral Therapy and Experimental Psychiatry* 3 (1972): 23–30.

124 *sex with both men and women*: Kent C. Berridge, "Pleasures of the brain," *Brain and Cognition* 52, no. 1 (2003): 106–28, doi:10.1016/S0278-2626(03)00014-9.

125 *what caused it*: Kent C. Berridge, Isabel L. Venier, and Terry E. Robinson, "Taste reactivity analysis of 6-hydroxydopamine-induced aphagia: Implications for arousal and anhedonia hypotheses of dopamine function," *Behavioral Neuroscience* 103, no. 1 (1989): 36–45.
To test whether dopamine caused pleasure, Berridge returned to the rodent smile, beginning a strange debate about the inner life of a rat. Roy Wise believed the rats could not possibly feel pleasure without dopamine and that their smiles were a reflex, their brain and muscles carrying out programming in response to a stimulus, with no conscious feeling of gratification. He had a point. Like bitterness, a sweet taste evokes an automatic

reaction: newborn babies smile when sugar is placed on their lips; so do animals with most of their brains removed. But Berridge hypothesized the rat smiles were exactly what they appeared to be: a genuine expression of satisfaction—just caused by something other than dopamine.

He hatched a clever experiment. Anyone who has fallen ill while eating finds the food that made them sick becomes persistently disgusting. This is a learned behavior. If Berridge could do the same for rats, changing their smiles to frowns, it would demonstrate the expressions were not lobotomized reflexes—which resist conditioning—but the real thing. He gave rats a dopamine blocker and a drug that caused nausea, followed by sips of sweetened water. Afterward, they all gaped with distaste at sugar water—now they hated it.

125 *directly cause pleasure*: Susana Peciña and Kent C. Berridge, "Opioid site in nucleus accumbens shell mediates eating and hedonic 'liking' for food: Map based on microinjection Fos plumes," *Brain Research* 863, nos. 1–2 (2000): 71–86.

127 *over the course of a lifetime*: Wolfram Schultz, Peter Dayan, and P. Read Montague, "A neural substrate of prediction and reward," *Science* 275, no. 5306: 1593–99, doi:10.1126/science.275.5306.1593; Wolfram Schultz, "The reward signal of midbrain dopamine neurons," *News in Physiological Science* 14 (1999): 67–71.

128 *maybe even happiness itself*: Morten L. Kringelbach and Kent C. Berridge, "The Neurobiology of Pleasure and Happiness," in *Oxford Handbook of Neuroethics*, eds. Judy Illes and Barbara J. Sahakian (Oxford, UK: Oxford University Press, 2011), 15–32.

129 *sipping sugar water*: Ivan E. de Araujo, Albino J. Oliveira-Maia, Tatyana D. Sotnikova, Raul R. Gainetdinov, Marc G. Caron, Miguel A. L. Nicolelis, and Sidney A. Simon, "Food reward in the absence of taste receptor signaling," *Neuron* 57, no. 6 (2008): 930–41, doi:10.1016/j.neuron.2008.01.032.

131 *into an insecticide*: Walter Gratzer, "Light on Sweetness: the Discovery of Aspartame," in *Eurekas and Euphorias: The Oxford Book of Scientific Anecdotes* (Oxford, UK: Oxford University Press, 2004), 32.

132 *contribute to diabetes*: Jotham Suez, Tal Korem, David Zeevi, Gili Zilberman-Schapira, Christoph A. Thaiss, Ori Maza, David Israeli, Niv Zmora, Shlomit Gilad, Adina Weinberger, Yael Kuperman, Alon Harmelin, Ilana Kolodkin-Gal, Hagit Shapiro,

Zamir Halpern, Eran Segal, and Eran Elinav, "Artificial sweeteners induce glucose intolerance by altering the gut microbiota," *Nature* 514 (October 2014): 181–86, doi:10.1038/nature13793.

132 *has a bitter edge*: Caroline Hellfritsch, Anne Brockhoff, Frauke Stähler, Wolfgang Meyerhof, and Thomas Hofmann, "Human psychometric and taste receptor responses to steviol glycosides," *Journal of Agricultural and Food Chemistry* 60, no. 27 (2012): 6782–93.

Chapter 6: Gusto and Disgust

136 *butchered for the choicest parts*: Charles Darwin, *The Voyage of the Beagle* (New York: P. F. Collier and Son, 1909), 86, http://www1 .umassd.edu/specialprograms/caboverde/darwin.html.

136 *practiced cannibalism*: Ann Chapman, *European Encounters with the Yahgan People of Cape Horn, Before and After Darwin* (New York: Cambridge University Press, 2010), 180.

142 *"something which smells bad"*: Paul Ekman and Wallace Friesen, "Constants across cultures in the face and emotion," *Journal of Personality and Social Psychology* 17, no. 2 (1971): 124–29.

143 *larger groups than other primates do*: Seth D. Dobson and Chet C. Sherwood, "Correlated evolution of brain regions involved in producing and processing facial expressions in anthropoid primates," *Biology Letters* 7, no. 1 (2010): 86–88, doi:10.1098 /rsbl.2010.0427.

143 *precise forms of communication*: For a discussion of the evolution of language, gesture, and facial expression, see Maurizio Gentilucci and Michael C. Corballis, "The Hominid that Talked," in *What Makes Us Human*, ed. Charles Pasternak (Oxford, UK: Oneworld Publications, 2007), 49–70.

144 *respond with heightened alertness*: Daniel M. T. Fessler, Serena J. Eng, and C. David Navarrete, "Elevated disgust sensitivity in the first trimester of pregnancy: Evidence supporting the compensatory prophylaxis hypothesis," *Evolution and Human Behavior* 26, no. 4 (2005): 344–51, doi:10.1016/j.evolhumbehav.2004.12.001.

145 *endless, changing threats*: Valerie Curtis, Robert Aunger, and Tamer Rabie, "Evidence that disgust evolved to protect from risk of disease," supplement, *Proceedings of the Royal Society B: Biological Sciences* 271 (2004): S131–33, doi:10.1098/rsbl.2003.0144; Valerie

Curtis, "Why disgust matters," *Philosophical Transactions of the Royal Society B: Biological Sciences* 366, no. 1583 (2011): 3478–90, doi:10.1098/rstb.2011.0165; Valerie Curtis, "Dirt, disgust and disease: A natural history of hygiene," *Journal of Epidemiology and Community Health* 61, no. 8 (2007): 660–64, doi:10.1136/jech.2007.062380; Valerie Curtis, Mícheál de Barra, and Robert Aunger, "Disgust as an adaptive system for disease avoidance behaviour," *Philosophical Transactions of the Royal Society B: Biological Sciences* 366, no. 1563 (2011): 389–401, doi:10.1098/rstb.2010.0117.

146 *true feelings to the outside world*: Ralph Adolphs, Daniel Tranel, Michael Koenigs, and Antonio R. Damasio, "Preferring one taste over another without recognizing either," *Nature Neuroscience* 8, no. 7 (2005): 860–61, doi:10.1038/nn1489.

147 *labeled the food "delicious"*: Ralph Adolphs, "Dissociable neural systems for recognizing emotions," *Brain and Cognition* 52, no. 1 (2003): 61–69, doi:10.1016/S0278-2626(03)00009-5.

147 *empathetic responses unite*: Bruno Wicker, Christian Keysers, Jane Plailly, Jean-Pierre Royet, Vittorio Gallese, and Giacomo Rizzolatti, "Both of us disgusted in my insula: The common neural basis of seeing and feeling disgust," *Neuron* 40, no. 3 (2003): 655–64, http://www.ncbi.nlm.nih.gov/pubmed/14642287.

147 *the brighter the insula burns*: See, for example, Mbemba Jabbe, Marte Swart, and Christian Keysers, "Empathy for positive and negative emotions in the gustatory cortex," *NeuroImage* 34, no. 4 (2008): 1744–53, doi:10.1016/j.neuroimage.2006.10.032.

148 *relationships and social personae*: For some discussion, see A. D. (Bud) Craig, *"How do you feel—now?"* 59–70; Isabella Mutschler, Céline Reinbold, Johanna Wanker, Erich Seifritz, and Tonio Ball, "Structural basis of empathy and the domain general region in the anterior insular cortex," *Frontiers in Human Neuroscience* 7: 177, doi:10.3389/fnhum.2013.00177; James Woodward and John Allman, "Moral intuition: Its neural substrates and normative significance," *Journal of Physiology–Paris* 101, nos. 4–6 (2007): 179–202.

148 *primitive form of morality*: H. A. Chapman, D. A. Kim, J. M. Susskind, and A. K. Anderson, "In bad taste: evidence for the oral origins of moral disgust," *Science* 323, no. 5918 (2009): 1222–26, doi:10.1126/science.1165565.

NOTES

149 *nearly half of adults did*: Paul Rozin, April Fallon, and MaryLynn Augustoni-Ziskind, "The child's conception of food: The development of contamination sensitivity to 'disgusting' substances," *Developmental Psychology* 21, no. 6: 1075–79, doi:10.1037//0012 -1649.21.6.1075.

154 *hairy beast*: Nick Hazelwood, *Savage: The Life and Times of Jemmy Button* (New York: St. Martin's Press, 2000), 338.

154 *"Teheran ape-child"*: Lucien Malson, *Wolf Children and the Problem of Human Nature* (New York: Monthly Review Press, 1972). Also contains the text of Itard's "The Wild Boy of Aveyron."

157 *Fame shopwindow*: Laudan, *Cuisine and Empire*, location 295.

159 *predictable and reliable*: William H. Brock, *Justus von Liebig: The Chemical Gatekeeper* (Cambridge, UK: Cambridge University Press, 1997), 216–29.

Chapter 7: Quest for Fire

163 *more vivid and pleasurable*: McGee, *On Food and Cooking*, 394–95.

163 *obscures these sensations*: Bernd Nilius and Giovanni Appendino, "Tasty and healthy TR(i)Ps: The human quest for culinary pungency," *EMBO Reports* 12, no. 11 (2011): 1094–101, doi:10.1038 /embor.2011.200.

165 *bland chilies than to hot ones*: David C. Haak, Leslie A. McGinnis, Douglas J. Levey, and Joshua J. Tewksbury, "Why are not all chilies hot? A trade-off limits pungency," *Proceedings of the Royal Society B: Biological Sciences* 279 (2011): 2012–17, doi:10.1098 /rspb.2011.2091; Joshua J. Tewksbury, Karen M. Reagan, Noelle J. Machnicki, Tomas A. Carlo, David C. Haak, Alejandra Lorena Calderon Penaloza, and Douglas J. Levey, "Evolutionary ecology of pungency in wild chilies," *Proceedings of the National Academy of Sciences* 105, no. 33 (2008): 11808–11, doi:10.1073 /pnas.0802691105.

167 *jalapeño, ancho, serrano, and tabasco peppers*: Linda Perry and Kent V. Flannery, "Pre-Columbian use of chili peppers in the valley of Oaxaca, Mexico," *Proceedings of the National Academy of Sciences* 104, no. 29 (2007): 11905–9.

167 *ancient craze to rival the modern one*: Linda Perry, Ruth Dickau, Sonia Zarrillo, Irene Holst, Deborah Pearsall, Dolores R. Piperno,

NOTES

Richard G. Cooke, Kurt Rademaker, Anthony J. Ranere, J. Scott Raymond, Daniel H. Sandweiss, Franz Scaramelli, and James A. Zeidler, "Starch fossils and the domestication and dispersal of chili peppers (*Capsicum* spp. L.) in the Americas," *Science* 315, no. 5814 (2007): 986–88, doi:10.1126/science.1136914.

168 *"loaded each year with it"*: Christopher Columbus, *The Log of Christopher Columbus*, trans. Robert H. Fuson (Camden, ME: International Marine Publishing, 1987).

169 *the name "calicut" pepper*: Jean Andrews, *Peppers: The Domesticated Capsicums* (Austin: University of Texas Press, 1984), 5.

169 *ports of call around the world*: Michael Krondl, *The Taste of Conquest: The Rise and Fall of the Three Great Cities of Spice* (New York: Ballantine Books, 2007), 170.

169 *wrote a song*: Ibid., 172.

170 *25 times the size it was fifty years ago*: UN Food and Agriculture Organization data.

170 *That number has more than doubled*: USDA Economic Research Service data.

173 *it's broad and flat*: Paul Bosland, interview.

175 *chili burn was a form of pain*: T. S. Lee, "Physiological gustatory sweating in a warm climate," *Journal of Physiology* 124 (1954): 528–42.

176 *serve the emperor's court as eunuchs*: Arpad Szallasi and Peter M. Blumberg, "Vanilloid (capsaicin) receptors and mechanisms," *Pharmacological Reviews* 51, no. 2 (1999): 159–212; Mary M. Anderson, *Hidden Power: The Palace Eunuchs of Imperial China* (Buffalo, NY: Prometheus, 1990), 15–18 and 307–11.

176 *"(available at a low price)"*: Sigmund Freud, *Cocaine Papers*, ed. Robert Byck (New York: Plume, 1975), 123.

177 *bodies literally overheated*: Narender R. Gavvaa, James J. S. Treanor, Andras Garami, Liang Fang, Sekhar Surapaneni, Anna Akrami, Francisco Alvarez, Annette Bake, Mary Darling, Anu Gore, Graham R. Jang, James P. Kesslak, Liyun Ni, Mark H. Norman, Gabrielle Palluconi, Mark J. Rose, Margaret Salfi, Edward Tan, Andrej A. Romanovsky, Christopher Banfield, and Gudarz Davar, "Pharmacological blockade of the vanilloid receptor TRPV1 elicits marked hyperthermia in humans," *Pain* 136, nos. 1–2 (2008): 202–10, doi:10.1016/j.pain.2008.01.024.

179 *a heat receptor*: Arpad Szallasi, "The vanilloid (capsaicin) receptor:

267

Receptor types and species specificity," *General Pharmacology* 25 (1994): 223–43.

181 *soldiers in the Himalayas*: Sudha Ramachandran, "Indian Defense Spices Things Up," *Asia Times Online*, July 8, 2009, http://www.atimes.com/atimes/South_Asia/KG08Df01.html.

181 *precursor to diabetes*: Celine E. Riera, Mark O. Huising, Patricia Follett, Mathias Leblanc, Jonathan Halloran, Roger Van Andel, Carlos Daniel de Magalhaes Filho, Carsten Merkwirth, and Andrew Dillin, "TRPV1 pain receptors regulate longevity and metabolism by neuropeptide signaling," *Cell* 157, no. 5 (2014): 1023–36, doi:10.1016/j.cell.2014.03.051.

181 *but sometimes they die*: Peter Holzer, "The pharmacological challenge to tame the transient receptor potential vanilloid-1 (TRPV1) nocisensor," *British Journal of Pharmacology* 155, no. 8 (2008): 1145–62, doi:10.1038/bjp.2008.351; Peter Holzer interview, March 2012.

182 *raises metabolic rates*: Keith Singletary, "Red Pepper: Overview of potential health benefits," *Nutrition Today* 46, no. 1 (2011): 33–47.

182 *the slaking of thirst*: R. Eccles, L. Du-Plessis, Y. Dommels, and J. E. Wilkinson, "Cold pleasure: Why we like ice drinks, ice-lollies and ice cream," *Appetite* 71 (2013): 357–60, doi:10.1016/j.appet.2013.09.011.

183 *always chose the mild cracker first*: Paul Rozin and Deborah Schiller, "The nature and acquisition of a preference for chili pepper by humans," *Motivation and Emotion* 4, no. 1 (1980): 77–101.

184 *was razor-thin*: Ibid., 97.

185 *the two closely overlap*: Siri Leknes and Irene Tracey, "A common neurobiology for pain and pleasure," *Nature Reviews: Neuroscience* 9, no. 4 (2008): 314–20, doi:10.1038/nrn2333.

186 *weren't expecting the pain to end*: Siri Leknes, Michael Lee, Chantal Berna, Jesper Andersson, and Irene Tracey, "Relief as a reward: Hedonic and neural responses to safety from pain," *PloS One* 6, no. 4 (2011): e17870, doi:10.1371/journal.pone.0017870.

Chapter 8: The Great Bombardment

190 *mispronunciation of the word*: "Tayto's Place in World History," *The Independent*, May 6, 2006, http://www.independent.ie/unsorted/features/taytos-place-in-world-history-26383239.html.

190 *around the same time*: Herr's company website, http://www.herrs.com; Frito-Lay history on Funding Universe website, http://www.funding universe.com/company-histories/frito-lay-company-history/.

191 *a sense of tradition and ritual*: Laudan, *Cuisine and Empire*, location 958.

192 *potatoes, sugary beverages, and red meat*: Dariush Mozaffarian, Tao Hao, Eric B. Rimm, Walter C. Willett, and Frank B. Hu, "Changes in diet and lifestyle and long-term weight gain in women and men," *The New England Journal of Medicine* 364, no. 25 (2011): 2392–404, doi:10.1056/NEJMoa1014296.

193 *Chilies were often used to flavor it*: Ellen Messer, "Potatoes (White)," chapter II. B.3 in *Cambridge World History of Food*, eds. Kenneth F. Kiple and Kriemhild Coneè Ornelas, http://www.cambridge.org /us/books/kiple/potatoes.htm.

194 *somebody's back room, garage, or barn*: Dirk Burhans, *Crunch!: A History of the Great American Potato Chip* (Madison, WI: Terrace Books, 2008), Kindle location 322.

194 *powerful rush to the brain's pleasure centers*: Kent C. Berridge, "The debate over dopamine's role in reward: The case for incentive salience," *Psychopharmacology* 191, no. 3 (2007): 391–431, doi:10.1007/s00213-006-0578-x.

195 *attention became more focused and acute*: Clare E. Turner, Winston D. Byblow, Cathy M. Stinear, and Nicholas R. Gant, "Carbohydrate in the mouth enhances activation of brain circuitry involved in motor performance and sensory perception," *Appetite* 80 (2014): 212–19, doi:10.1016/j.appet.2014.05.020.

195 *the richer it tastes*: Marta Yanina Pepino, Latisha Love-Gregory, Samuel Klein, and Nada A. Abumrad, "The fatty acid translocase gene, CD36, and lingual lipase influence oral sensitivity to fat in obese subjects," *Journal of Lipid Research* 53, no. 3 (2012): 561–66, doi:10.1194/jlr.M021873.

196 *they loved it*: Amy J. Tindell, Kyle S. Smith, Susana Peciña, Kent C. Berridge, and J. Wayne Aldridge, "Ventral pallidum firing codes hedonic reward: When a bad taste turns good," *Journal of Neurophysiology* 96, no. 5 (2006): 2399–409, doi:10.1152/jn.00576.2006.

196 *fusion of these two bad tastes*: Yuki Oka, Matthew Butnaru, Lars von Buchholtz, Nicholas J. P. Ryba, and Charles S. Zuker, "High salt recruits aversive taste pathways," *Nature* 494 (2013): 472–75, doi:10.1038/nature11905.

197 *literally addicted to salt*: Michael J. Morris, Elisa S. Na, and Alan Kim Johnson, "Salt craving: The psychobiology of pathogenic sodium intake," *Physiology & Behavior* 94, no. 5 (2008): 709–21, doi:10.1016/j.physbeh.2008.04.008.

198 *over time, they gained weight*: Jacques Le Magnen, *Hunger* (Cambridge, UK: Cambridge University Press, 1985), 42.

198 *fights, assaults, and disorderly conduct*: Eliza Barclay, "Food As Punishment: Giving US Inmates 'The Loaf' Persists," *Morning Edition*, NPR, January 2, 2014, http://www.npr.org/blogs/the salt/2014/01/02/256605441/punishing-inmates-with-the-loaf -persists-in-the-u-s.

199 *keeping it alive for others*: Barbara J. Rolls, Edmund T. Rolls, Edward A. Rowe, and Kevin Sweeney, "Sensory specific satiety in man," *Physiology & Behavior* 27 (1980): 137–42.

200 *"ham and motherfuckers"*: Robert E. Peavey, *Praying for Slack: A Marine Corps Tank Commander in Vietnam* (Minneapolis: Zenith Imprint Press, 2004), 189.

203 *vision, memory, and knowledge*: Kathrin Ohla, Ulrike Toepel, Johannes le Coutre, and Julie Hudry, "Visual-gustatory interaction: Orbitofrontal and insular cortices mediate the effect of high-calorie visual food cues on taste pleasantness," *PloS One* 7, no. 3 (2012): e32434, doi:10.1371/journal.pone.0032434.

203 *makes pink yogurt taste worse*: Vanessa Harrar and Charles Spence, "The taste of cutlery: How the taste of food is affected by the weight, size, shape, and colour of the cutlery used to eat it," *Flavour* 2, no. 21 (2013), doi:10.1186/2044-7248-2-21.

204 *saltier from a blue bowl*: Charles Spence, Vanessa Harrar, and Betina Piqueras-Fiszman, "Assessing the impact of the tableware and other contextual variables on multisensory flavour perception," *Flavour* 1, no. 7 (2012), doi:10.1186/2044-7248-1-7.

204 *wine is more expensive, it tastes better*: Hilke Plassmann, John Doherty, Baba Shiv, and Antonio Rangel, "Marketing actions can modulate neural representations of experienced pleasantness," *Proceedings of the National Academy of Sciences* 105, no. 3 (2008): 1050–54.

204 *"chicory," "coal," and "musk"*: Gil Morrot, Frederic Brochet, and Denis Dubourdieu, "The color of odors," *Brain and Language* 79, no. 2 (2001): 309–20, doi:10.1006/brln.2001.2493.

205 *bringing experience to bear on flavor*: Samuel M. McClure, Jian Li, Damon Tomlin, Kim S. Cypert, Latane M. Montague, and P.

Read Montague, "Neural correlates of behavioral preference for culturally familiar drinks," *Neuron* 44, no. 2 (2004): 379–87, doi:10.1016/j.neuron.2004.09.019.

206 *The volume of choices was a deterrent*: Sheena S. Iyengar and Mark R. Lepper, "When choice is demotivating: Can one desire too much of a good thing?" *Journal of Personality and Social Psychology* 79, no. 6 (2000): 995–1006.

206 *during the moment of decision*: Hilke Plassmann, John O. Doherty, and Antonio Rangel, "Orbitofrontal cortex encodes willingness to pay in everyday economic transactions," *The Journal of Neuroscience* 27, no. 37 (2007): 9984–88, doi:10.1523/JNEUROSCI.2131-07.2007.

209 *on the website* Gawker: Hamilton Nolan, "Americans Will Be Drugged to Believe Their Soda Is Sweeter," *Gawker*, December 3, 2013, http://gawker.com/americans-will-be-drugged-to-believe -their-soda-is-swee-1475526047.

210 *into the right kind of muscle tissue*: Nicola Jones, "A taste of things to come? Researchers are sure that they can put lab-grown meat on the menu—if they can just get cultured muscle cells to bulk up," *Nature* 468 (2010): 752–53.

210 *"a bit like cake"*: Davide Castelvecchi, "Researchers Put Synthetic Meat to the Palate Test," *Nature News Blog*, August 15, 2013, http://blogs.nature.com/news/2013/08/researchers-put-synthetic -meat-to-the-palate-test.html.

211 *"meal in a glass"*: Rob Rhinehart, "How I Stopped Eating Food," *Mostly Harmless* (blog), February, 13 2013, http://robrhinehart .com/?p=298.

213 *share it with the world*: Nimesha Ranasinghe website, http://nime sha.info/projects.html; Nimesha Ranasinghe, Ryohei Nakatsu, Nii Hideaki, and Ponnampalam Gopalakrishnakone, "Tongue-mounted interface for digitally actuating the sense of taste," *Proceedings of the 16th IEEE International Symposium on Wearable Computers* (June 2012): 80–87, doi:10.1109/ISWC.2012.16, ISSN: 1550-4816; Nimesha Ranasinghe, Kasun Karunanayaka, Adrian David Cheok, O. N. N. Fernando, Hideaki Nii, Ponnampalam Gopalakrishnakone, "Digital Taste and Smell Communication," *Proceedings of International Conference on Body Area Networks, BodyNets 2011* (November 2011): 78–84; Nimesha Ranasinghe, A. D. Cheok, O. N. N. Fernando, H. Nii, and G. Ponnampalam, "Electronic taste stimulation," *Proceedings of the 13th*

International Conference on Ubiquitous Computing (2011): 561–62, doi:10.1145/2030112.2030213.

Chapter 9: The DNA of Deliciousness

215 *a strong umami signature*: McGee, *On Food and Cooking*, 237.

218 *had both been wrong*: René Dubos, *Louis Pasteur: Free Lance of Science* (Boston: Little, Brown and Company, 1950), 41, 116–34.

220 *the skin remained crispy*: Hervé This, "Modelling dishes and exploring culinary 'precisions': The two issues of molecular gastronomy," supplement, *British Journal of Nutrition* 93, no. 1 (2007): S139–S146, doi:10.1079/BJN20041352.

221 *"new techniques and dishes"*: "Cooking Statement," The Fat Duck website, http://www.thefatduck.co.uk/Heston-Blumenthal/Cooking -Statement/.

224 *"traditional fermentative processes"*: Daniel Felder, Daniel Burns, and David Chang, "Defining microbial terroir: The use of native fungi for the study of traditional fermentative processes," *International Journal of Gastronomy and Food Science* 1, no. 1 (2011): 64–69, doi:10.1016/j.ijgfs.2011.11.003.

230 *"what goes on inside our soufflés"*: Leo Hickman, "Doctor Food," *The Guardian*, April 19, 2005, http://www.theguardian.com /news/2005/apr/20/food.science.

232 *Western Europe and North America*: Yong-Yeol Ahn, Sebastian E. Ahnert, James P. Bagrow, and Albert-Laszlo Barabasi, "Flavor network and the principles of food pairing," *Scientific Reports* 196, no. 1 (2011): 1–7, doi:10.1038/srep00196.

232 *foie gras with jasmine sauce*: Hickman, "Doctor Food."

234 *forged a creative bond*: Chris Nay, "When Machines Get Creative: The Virtual Chef," *Building a Smarter Planet* (blog), December 12, 2013, http://asmarterplanet.com/blog/2013/12/virtualchef.html. "Cognitive Cookbook," IBM website, http://www.ibm.com/smarter planet/us/en/cognitivecooking/food.html.

235 *set out to isolate it*: Kenzo Kurihara, "Glutamate: From discovery as a food flavor to role as a basic taste (umami)," *American Journal of Clinical Nutrition* 90, no. 3 (2009): 719S–722S, doi: 10.3945 /ajcn.2009.27462D.

243 *grapes in Bordeaux*: Gregory V. Jones, "Climate change: Obser-

vations, projections and general implications for viticulture and wine production," Whitman College Economics Department Working Paper, 2007.

243 *for more global warming*: John McQuaid, "What Rising Temperatures May Mean for World's Wine Industry," *Yale Environment 360*, December 19, 2011, http://e360.yale.edu/feature/what _global_warming_may_mean_for_worlds_wine_industry/2478/.

Bibliography

Allman, John Morgan. *Evolving Brains.* New York: Scientific American Library, 2000.

Andrews, Jean. *Peppers: The Domesticated Capsicums.* Austin: University of Texas Press, 1984.

Baker, Phil. *The Book of Absinthe: A Cultural History.* New York: Grove Press, 2007.

Boring, Edwin G. *Sensation and Perception in the History of Experimental Psychology.* New York: Appleton-Century-Crofts, Inc., 1942.

Brillat-Savarin, Jean Anthelme. *The Physiology of Taste: Or Meditations on Transcendental Gastronomy.* Translated by M. F. K. Fisher. New York: Vintage electronic edition, 2009.

Brock, William H. *Justus von Liebig: The Chemical Gatekeeper.* Cambridge, UK: Cambridge University Press, 1997.

Burhans, Dirk. *Crunch!: A History of the Great American Potato Chip.* Madison, WI: Terrace Books, 2008.

Carterette, Edward C. and Morton P. Friedman, eds. *Handbook of Perception, Volume VIA: Tasting and Smelling.* New York: Academic Press, 1978.

Cavalli-Sforza, L. Luca, Paolo Menozzi, and Alberto Piazza. *The History and Geography of Human Genes.* Princeton: Princeton University Press, 1994.

Cervantes, Miguel de. *Don Quixote.* Translated by Edith Grossman. New York: HarperCollins, 2009.

Chapman, Ann. *European Encounters with the Yahgan People of Cape Horn, Before and After Darwin.* New York: Cambridge University Press, 2010.

Columbus, Christopher. *The Log of Christopher Columbus.* Translated by

BIBLIOGRAPHY

Robert H. Fuson. Camden, ME: International Marine Publishing, 1987.

Cowell, E. B., F. Max Muller, and J. Takakusu, translators. *Buddhist Mahāyāna Texts*. New York: Dover Publications, 1969.

Darwin, Charles. *The Expression of Emotions in Man and Animals*. New York: D. Appleton and Co., 1899. Accessed via Project Gutenberg. http://www.gutenberg.org/files/1227/1227-h/1227-h.htm.

———. *The Voyage of the Beagle*. New York: P. F. Collier and Son, 1909. Accessed via Internet Wiretap. http://www1.umassd.edu/special programs/caboverde/darwin.html.

Dubos, René. *Louis Pasteur: Free Lance of Science*. Boston: Little, Brown and Company, 1950. Accessed via University of California Digital Library. https://archive.org/details/louispasteurfree009068mbp.

Ekman, Paul, ed. *Darwin and Facial Expression: A Century of Research in Review*. Los Altos, CA: Malor Books, 2006.

Finger, Stanley. *Origins of Neuroscience: A History of Explorations into Brain Function*. Oxford, UK: Oxford University Press, 2001.

Freud, Sigmund. *Cocaine Papers*. Robert Byck, editor. New York: Plume, 1975.

Galloway, J. H. *The Sugar Cane Industry: An Historical Geography from its Origins to 1914*. Cambridge, UK: Cambridge University Press, 1989.

Gopnik, Alison, Andrew N. Meltzoff, and Patricia K. Kuhl. *The Scientist in the Crib: What Early Learning Tells Us About the Mind*. New York: HarperCollins, 2000.

Gratzer, Walter. *Eurekas and Euphorias: The Oxford Book of Scientific Anecdotes*. Oxford, UK: Oxford University Press, 2004.

Hazelwood, Nick. *Savage: The Life and Times of Jemmy Button*. New York: St. Martin's Press, 2000.

Herz, Rachel. *That's Disgusting: Unraveling the Mysteries of Repulsion*. New York: W. W. Norton, 2012.

Heyn, Birgit. *Ayurveda: The Indian Art of Natural Medicine and Life Extension*. Rochester, VT: Healing Arts Press, 1990.

Homer. *The Odyssey*. Translated by Robert Fagles. New York: Penguin Classics, 2002.

Hounshell, David A. and John Kenly Smith, Jr. *Science and Corporate Strategy: DuPont R&D 1902–1980*. In *Studies in Economic History and Policy: The United States in the Twentieth Century*. Cambridge, UK: Cambridge University Press, 1988.

BIBLIOGRAPHY

Illes, Judy, and Barbara J. Sahakian, eds. *Oxford Handbook of Neuroethics*. Oxford, UK: Oxford University Press, 2011.

James, Henry, ed. *Letters of William James*. Boston: Atlantic Monthly Press, 1920.

Kelley, Patricia H., Michal Kowalewski, and Thor A. Hansen, eds. *Predator-Prey Interactions in the Fossil Record*. New York: Kluwer Academic/Plenum Publishers, 2003.

Keysers, Christian. *The Empathic Brain: How the Discovery of Mirror Neurons Changes Our Understanding of Human Nature*. Groningen, Netherlands: Social Brain Press, 2011.

Kieschnick, John. *The Impact of Buddhism on Chinese Material Culture*. Princeton, NJ: Princeton University Press, 2003.

Kinstedt, Paul S. *Cheese and Culture: A History of Cheese and Its Place in Western Civilization*. White River Junction, VT: Chelsea Green, 2012.

Korsmeyer, Carolyn. *Making Sense of Taste: Food and Philosophy*. Ithaca, NY: Cornell University Press, 1999.

Kringelbach, Morton L. *The Pleasure Center: Trust Your Animal Instincts*. Oxford, UK: Oxford University Press, 2009.

Krondl, Michael. *The Taste of Conquest: The Rise and Fall of the Three Great Cities of Spice*. New York: Ballantine Books, 2007.

Laudan, Rachel. *Cuisine and Empire: Cooking in World History*. Berkeley: University of California Press, 2013.

Le Magnen, Jacques. *Problems in the Behavioural Sciences*. Bk. 3, *Hunger*. Cambridge, UK: Cambridge University Press, 1985.

Lieberman, Daniel E. *The Evolution of the Human Head*. Cambridge, MA: The Belknap Press of Harvard University Press, 2011.

Malmberg, Annika B. and Keith R. Bley, eds. *Turning Up the Heat on Pain: TRPV1 Receptors in Pain and Inflammation*. Boston: Birkhauser Verlag, 2005.

Malson, Lucien. *Wolf Children and the Problem of Human Nature*. New York: Monthly Review Press, 1972.

McGee, Harold. *On Food and Cooking: The Science and Lore of the Kitchen*. New York: Scribner, 2004.

McGovern, Patrick. *Uncorking the Past: The Quest for Wine, Beer and Other Alcoholic Beverages*. Berkeley: University of California Press, 2009.

McSweeney, P. L. H., ed. *Cheese Problems Solved*. Cambridge, UK: Woodhead Publishing Ltd., 2007.

Mintz, Sidney. *Sweetness and Power: The Place of Sugar in Modern History*. New York: Penguin Books, 1985.

BIBLIOGRAPHY

Moss, Michael. *Salt Sugar Fat: How the Food Giants Hooked Us.* New York: Random House, 2013.

Newton, Michael. *Savage Girls and Wild Boys: A History of Feral Children.* New York: Picador, 2002.

O'Connell, Sanjida. *Sugar: The Grass That Changed the World.* London: Virgin Books, 2004.

Ostwald, C. W. W. *An Introduction to Theoretical and Applied Colloid Chemistry: The World of Neglected Dimensions.* New York: John Wiley & Sons, 1917. Accessed via University of California Digital Library. https://archive.org/details/theoapplicolloid00ostwrich.

The Oxford English Dictionary, compact edition. Oxford, UK: Oxford University Press, 1980.

Parker, Matthew. *The Sugar Barons: Family, Corruption, Empire, and War in the West Indies.* New York: Walker, 2011.

Pasternak, Charles, ed. *What Makes Us Human.* Oxford, UK: Oneworld Publications, 2007.

Peavey, Robert E. *Praying for Slack: A Marine Corps Tank Commander in Vietnam.* Minneapolis: Zenith Imprint Press, 2004.

Plato. *Timaeus.* Translated by Benjamin Jowett. Accessed via MIT Internet Classics Archive. http://classics.mit.edu/Plato/timaeus.html.

Prescott, John and Beverly Tepper, eds. *Genetic Variation in Taste Sensitivity.* New York: Marcel Dekker, 2004.

Reston Jr., James. *Warriors of God: Richard the Lionheart and Saladin in the Third Crusade.* New York: Anchor Books, 2007.

Richardson, Tim. *Sweets: A History of Candy.* New York: Bloomsbury, 2002.

Savage-Rumbaugh, E. Sue, and Roger Lewin. *Kanzi: The Ape at the Brink of the Human Mind.* New York: John Wiley & Sons, 1994.

Shepherd, Gordon M. *Neurogastronomy: How the Brain Creates Flavor and Why It Matters.* New York: Columbia University Press, 2012.

Siegel, Ronald K. *Intoxication: The Universal Drive for Mind-Altering Substances.* New York: Park Street Press, 2011.

Stuart, Tristram. *The Bloodless Revolution: A Cultural History of Vegetarianism from 1600 to Modern Times.* New York: W. W. Norton, 2006.

This, Hervé. *Molecular Gastronomy: Exploring the Science of Flavor.* Translated by Malcolm DeBevoise. New York: Columbia University Press, 2006.

Wrangham, Richard. *Catching Fire: How Cooking Made Us Human.* New York: Basic Books, 2009.

Index

INDEX

Auckland, University of, 194–95
Aurochs, 86, 87
Australia, 109, 162, 191, 244
Australopithecines, 35–37
Austria, 156
Aymara people, 72, 183
Ayurveda, 8
Aztecs, 50, 176

Bab el-Mandeb Strait, 63–64, 72
Bacteria, 19, 49, 70, 143, 218, 222
  in fermentation process, 78,
    87–88, 224–29
  See also specific species
Badiano, Juan, 176
Badiano Codex (Cruz and Badiano),
  176, 177
Baghdad, 112–13
Bahamas, 167, 168
Barale, Roberto, 69–70
Barbados, 115–16
Barro Colorado Island, 81
Bartoshuk, Linda, 3, 68–69
Bavaria, 84
Baylor University, 204
BBC, 144
Beagle (ship), 135–37, 157
Beer, 78, 84, 89, 98, 109, 192
  bitterness of, 44, 50, 58, 66
Beethoven, Ludwig van, 130
Behavioral immune system, 145, 154
Behaviorism, 120–21
Belgium, 232
Benedictine order, 113
Berger, Lee, 35
Berger, Matthew, 35
Berridge, Kent, 122–28, 196, 262–63n
Bhallika, 111
Bible, 50, 153, 197
Birds, 18, 29, 31, 165
Bitterness, 2–3, 11–12, 28, 44,
    47–73, 117, 132, 163, 228,
    234, 241
  adaptation to, 72–73, 182–83
  of alcoholic beverages, 50, 58, 75,
    77, 85, 98–99

aversion to, 49, 139, 148, 207
combined with other flavors,
  50–53, 196
of plants, 37, 164
in prehistoric human diet, 37, 39
receptors for, 60–61, 70–71, 95,
  98, 131, 196, 207–8
variations in sensitivity to, 34,
  47–49, 53–59, 61–63, 65–69
virtual, 213
Bizarre Foods (TV show), 151
Blood sugar, 72, 181–82
Bloomberg, Michael, 108
Blumenthal, Heston, 220, 232
Bolivia, 72, 163, 165, 183
Bollongino, Ruth, 86
Bonaparte, Napoleon, 118, 156, 157
Bonobos, 39–40
Bordeaux, University of, 204
Boring, Edwin Garrigues, 1–5, 200
Borneo, 142
Bourdain, Anthony, 151
Brain stem, 42
Brassica, 47, 72
Brazil, 141
Brickley, Zoe, 227–28
Brillat-Savarin, Anthelme, 90–91
Brin, Sergey, 209
Britain, 66, 189–191, 209. *See also*
  England
British Empire, 115, 137, 142, 154
  navy of, 157
Broccoli, 47–48, 58, 61, 72, 164
Buddhism, 110–12, 128
Burke, Seamus, 190
Burma, 169
Burns, Daniel, 216–17
Bush, Barbara, 48
Bush, George H. W., 47, 72
Bush, George W., 48
Butabushi, 216–17, 222–23
Butyric acid, 93

Cacao beans, 50
Caffeine, 51, 72, 164
Calabria, 70–71

INDEX

INDEX

INDEX

INDEX

INDEX

Inuit, 66
Iowa, University of, 197
Iowa Primate Learning Sanctuary, 39
Iran, 65, 86, 112
Ireland, 189–90, 193
Islam, 112
Isothiocyanates, 72
Israel, 32, 64
Italy, 70, 77, 80, 156
    vineyards in, 80
Itard, Jean-Marc-Gaspard, 155
Iyengar, Sheena, 205

Jains, 111
James, William, 119–20
Japan, 89, 141, 142, 209, 234–35
    cuisine of, 215, 216, 222–24, 237–38
Jasper Hill Farm, 225–29
Jawless fish, 21–23, 61
Jeopardy! (TV show), 233
Jerusalem, 113
Jesus, 197
Jews, 50
    dietary laws of, 153
Jiahu (China), 82–85, 99
Johns Hopkins University, 131
Johnson, Alan Kim, 197
Juba, King, 178
Junk food, 191, 192, 212

Kant, Immanuel, 5–6
Kanzi, 39–40
Katsuobushi, 215–16, 222, 235
Kautilya, 111
Kehler, Andy and Mateo, 228–29
Keller, Thomas, 220
Kenya, 38, 67
Keysers, Christian, 11
Kickstarter, 212
Kilimanjaro, Mount, 45
Kitab al-Tabikh, 112
Klee, Harry, 242
Koji, 85, 89–90, 222
Kurti, Nicholas, 219, 230

Kuru, 141
Kuyavia (Poland), 88

Lactobacillus, 87, 88, 151
Lactose, 87–88, 225
Lahousse, Bernard, 232–33
Lake Assal, 44–45
Lalueza-Fox, Carles, 63
Landrau, Jacob, 50–52
Laudan, Rachel, 191
Lay's potato chips, 190
Leakey, Louis and Mary, 38
Le Coutre, Johannes, 203
Lee, T. S., 174–75, 179
Leknes, Siri, 186
Le Magnen, Jacques, 197–98
Lepper, Mark, 205
Liebig, Justus von, 157–58
Lille, University of, 217
Linemeyer, David, 208
Linnaeus, Carl, 8, 114
London School of Hygiene and Tropical Medicine, 143
Louis XVI, King of France, 193
Love and Death (film), 20
"Lucy," 36

Mad cow disease, 151
Magnetic resonance imaging (MRI), 11
    functional. *See* Functional MRI (fMRI) scans
Maillard reactions, 40, 158, 197, 230
Malapa cave (South Africa), 36
Malaysia, 80
Malson, Lucien, 154
Mammals, 87, 143, 165, 197
    evolution of, 24–27, 29, 31, 34, 42, 164
    See also Primates
Marie Antoinette, Queen of France, 193
Martire d'Anghiera, Pietro, 169
Mason, Arnold, 57
Massachusetts Institute of Technology, 59

INDEX

Neurons, 15, 73, 92, 95–97, 102,
    199, 206
  dopamine and, 123, 126–28, 185
  facial-muscle, 143, 145, 147
  of hedonic hotspots, 126, 185,
    186
  mapping, 12, 244
  spindle, 147–48
  See also Receptors
Neurotransmitters, 70, 123–29,
    145, 165, 185
New Guinea, 109, 114, 140
New Mexico State University, Chili
    Pepper Institute, 172
New Orleans, 76
Newton, Isaac, 2, 7
New York City, 215, 223–24,
    237–38
  Institute of Culinary Education,
    234
  International Culinary Center, 240
  restaurants and bars in, 215,
    239–41
  sugar consumption in, 108
Nicotine, 164
Nightshade family, 164, 193
Nobel Prize, 9
Noller, Carl, 53
Nordic Food Lab (Copenhagen),
    243–44

Obesity, 108, 130, 150, 195
Odyssey (Homer), 102–3
Olds, James, 120–23
Olduvai Gorge, 38
Olfactory mechanisms, 23, 30, 42,
    59, 91–92, 94, 122. *See also*
    Smell, sense of
On the Origin of Species (Darwin), 138
Operculum, 95
Opertech Bio, 206, 208, 229
Opiates, 125
Optics, 7
Orbitofrontal cortex, 93, 96–98,
    101–2, 199, 203, 206
  damage to, 145–46

Orexin, 126
Organic chemistry, 157
Oxford University, 43, 94, 97, 186,
    199

Paleolithic diet, 211
Palmer, Kyle, 206–7
Panama, 81
Paris, 75, 86, 156, 157
Parke-Davis Company, 176, 177
Parkinson's disease, 123
Parmentier, Antoine-Augustin, 193
Passover, 50
Pasteur, Louis, 217–18
Pasteur, Marie, 218
Patagonia, 135
Pavlovian conditioning, 194
Penicillium roqueforti, 89
Penides, 114
Pennsylvania, University of, 67, 70,
    81, 148
Permian extinction, 24
Perry, Linda, 166–67
Persia, 112
Peru, 135, 163, 167
  ancient, 84
Phenylindanes, 51
Phenylthiocarbamide (PTC), 53–58,
    61, 62, 68, 72
Pheromones, 61, 69
Philippines, 169
Phrenology, 8
Physiology of Taste, The (Brillat-
    Savarin), 90–91
Phytoliths, 46
Pichia burtonii, 222–23, 227
Pickles, 50, 88, 106
Pinta (ship), 168
Pisa, University of, 69
Pittsburgh, University of, 4
Plantations, sugar, 115, 116, 118
Plants, 6, 20, 30, 33, 65, 89, 150,
    211, 232
  breeding and cultivation of. *See*
    Agriculture
  defenses developed by, 49, 164

287

INDEX

Sassanid dynasty, 112
Savoriness. *See* Umami
Schultz, Wolfram, 127
Science magazine, 55
Scientific method, 2, 155, 242
Scoville, Wilbur, 176–78
Scoville units, 161–63, 170, 172,
    176–78, 181, 185
Sea anemones, 49
Searle pharmaceutical company, 131
Secarz, Lior Lev, 237–39
Secor, Stephen, 41
Senegal, 43
Senomyx, 207–8
Sensation and Perception in the History
    of Experimental Psychology
    (Boring), 2
Sesquiterpenes, 90
Seven Years War, 193
Sex, 6, 21, 95, 119–20, 125, 176
    cravings for, 110
    deviant or inappropriate, disgust
        in response to, 139, 140
    and pleasure centers of brain, 122
Sheep, 78, 86–89, 103
Shepherd, Gordon, 92, 94
Shepherd-Barr, Kirsten, 94
Siam, 169
Siberia, volcanic eruptions in, 23–24
Sicily, 113
Siddhartha Gautama (Buddha), 110–
    11, 113, 118
Silk Road, 112, 238
Simner, Julia, 100–101
Singapore, National University of,
    174
6-n-propylthiouracil (PROP), 58
Skinner, B. F., 120–21, 207
Skinner boxes, 207
Slave trade, 115, 116, 169
Smell, sense of, 2, 48, 49, 80,
    90–96, 98–102, 164
    brain structures involved in, 12,
        92, 101–2, 127, 199
    of children, 14
    delicious, 209

disgust and, 139, 142, 151
    evolution of, 18, 22–23, 26–27,
        30, 34, 42, 79, 90
    memory and, 23, 94, 139
    receptors for, 9, 27, 42, 99
    sex hormones and, 198
    spices and, 238, 239
    umami and, 234
    See also Olfactory mechanisms
Smithsonian Institution, 166, 240
    Human Origins Program, 44
Solanaceae, 164, 193
Sonoran Desert, 18
Soranzo, Nicole, 66
South Africa, 35
South America, 65
Soy-based foods, 78, 89, 96, 232, 233
    umami in, 98, 235
Soylent, 211–13
Spain, 50, 63, 168–69, 191
    colonies of, 114–15, 176, 193
    cuisine of, 220
Spices, 105, 113, 168–70, 236–39
    acidic, 105
    See also Chili peppers
Spindle neurons, 147
Splenda, 131
Stanford University, 53, 205
Stem cells, 208–10
Stevia, 108
Stevia rebaudiana, 132
Still Life with Absinthe (van Gogh),
    75
Stone Age tribes, 141, 142
Streptococcus, 87
Studies on Fermentation (Pasteur), 218
Sucralose, 131–32
Sucrose, 118, 130, 131
Sugar, 3, 107–19, 121–24, 126–32,
    164, 189, 196, 200
    amino acids combined with. *See*
        Maillard reactions
    bitterness and, 50–52, 72
    in fermentation process, 69,
        80–82, 84–85, 89, 98, 217,
        253